Hans Jonas

Also Available from Bloomsbury

Lev Shestov: Philosopher of the Sleepless Night, Matthew Beaumont
Contradiction Set Free, Hermann Levin Goldschmidt
Hannah Arendt's Ethics, Deirdre Lauren Mahony
Another Finitude: Messianic Vitalism and Philosophy, Agata Bielik-Robson
Modernism between Benjamin and Goethe, Matthew Charles
Heine and Critical Theory, Willi Goetschel

Hans Jonas

Life, Technology and the Horizons of Responsibility

Lewis Coyne

BLOOMSBURY ACADEMIC
LONDON • NEW YORK • OXFORD • NEW DELHI • SYDNEY

BLOOMSBURY ACADEMIC
Bloomsbury Publishing Plc
50 Bedford Square, London, WC1B 3DP, UK
1385 Broadway, New York, NY 10018, USA
29 Earlsfort Terrace, Dublin 2, Ireland

BLOOMSBURY, BLOOMSBURY ACADEMIC and the Diana logo are trademarks of
Bloomsbury Publishing Plc

First published in Great Britain 2021
This paperback edition published in 2022

Copyright © Lewis Coyne, 2021

Lewis Coyne has asserted his right under the Copyright, Designs and
Patents Act, 1988, to be identified as Author of this work.

For legal purposes the Acknowledgements on p. x constitute an extension
of this copyright page.

Cover design by Charlotte Daniels
Cover image: View from a rooftop looking over the village of Pripyat, Ukraine, where many
Chernobyl power plant workers lived (© Hugh Mitton / Alamy Stock Photo)

All rights reserved. No part of this publication may be reproduced or transmitted
in any form or by any means, electronic or mechanical, including photocopying,
recording, or any information storage or retrieval system, without prior
permission in writing from the publishers.

Bloomsbury Publishing Plc does not have any control over, or responsibility for, any
third-party websites referred to or in this book. All internet addresses given in this
book were correct at the time of going to press. The author and publisher regret
any inconvenience caused if addresses have changed or sites have ceased to
exist, but can accept no responsibility for any such changes.

A catalogue record for this book is available from the British Library.

Library of Congress Cataloging-in-Publication Data
Names: Coyne, Lewis, author.
Title: Hans Jonas: life, technology and the horizons of responsibility / Lewis Coyne.
Description: London; New York: Bloomsbury Academic, 2020. |
Includes bibliographical references and index.
Identifiers: LCCN 2020025655 (print) | LCCN 2020025656 (ebook) | ISBN 9781350102392
(hardback) | ISBN 9781350102408 (ebook) | ISBN 9781350102415 (epub)
Subjects: LCSH: Jonas, Hans, 1903–1993.
Classification: LCC B3279.J664 C69 2020 (print) | LCC B3279.J664 (ebook) | DDC 193–dc23.
LC record available at https://lccn.loc.gov/2020025655
LC ebook record available at https://lccn.loc.gov/2020025656

ISBN: HB: 978-1-3501-0239-2
PB: 978-1-3502-1666-2
ePDF: 978-1-3501-0240-8
eBook: 978-1-3501-0241-5

Typeset by Deanta Global Publishing Services, Chennai, India

To find out more about our authors and books visit www.bloomsbury.com
and sign up for our newsletters.

To Beccy and Arthur

Contents

Preface	viii
Acknowledgements	x
Introduction	1
1 The gnosticism of modernity	9
2 The philosophy of life I: The organism	37
3 The philosophy of life II: The *scala naturae*	65
4 Values and the good	95
5 New dimensions of responsibility	117
6 The politics of nature	145
7 Towards a richer bioethics	173
Conclusion	199
Notes	209
References	221
Index	232

Preface

This book was first conceived of several years ago. While researching Martin Heidegger's connection to environmentalism, I happened upon the name of Hans Jonas, Heidegger's one-time student. Intrigued, I obtained a copy of Jonas's *The Imperative of Responsibility* and immediately sensed that I had found something of a hidden gem. Here was a philosopher who brought a deep understanding of intellectual history to bear on the most pressing issues of our time, becoming a public figure in his native Germany in the process, and Anglophone philosophers, it seemed – even those working in the relevant fields – had frequently neglected him. Before long I had read everything written by and on Jonas in the English language. Convinced that he deserved greater attention, the vague outline of a book that contributed to this effort formed in my mind.

I am happy to say that in the time since Jonas has indeed been the subject of increased scholarly interest. The eleven-volume critical edition of his collected works has been published, bringing further lecture courses and the manuscript of *Organism and Freedom* to light, while two monographs, a collection of essays and numerous journal articles have been added to the English-language secondary literature.[1] And yet in the English-speaking world a great deal remains to be learnt both from and about Jonas's philosophy. (His extensive religious scholarship, which I largely set aside in observation of disciplinary boundaries, is in any case fairly well known to theologians.)

This work attempts to carry out three tasks in particular. First, it seeks to demonstrate the systematic nature of Jonas's philosophical thought, which spans the following areas: phenomenology, philosophical anthropology, bioethics, environmental ethics, the philosophy of technology, political theory and the history of ideas. Clearly, a book of this length cannot hope to analyse Jonas's contributions to each in fine-grained detail. That could be the aim of future works; what is needed at present is a critical account of Jonas's systematic attempt to draw from the above sub-disciplines. Second, this book attempts to increase our understanding of Heidegger's influence on Jonas, as, although all commentators agree that Jonas's philosophy is deeply indebted to that of his teacher, the exact nature of this debt is still to be determined. Last but not least, it seeks to further account for the Aristotelian elements of Jonas's thought, and ultimately suggests that Aristotle's influence on Jonas is in fact second only to that of Heidegger. In pursuit of these aims, I draw on the full range of Jonas's philosophical output, including archival material and the few works yet to be translated from the German.

Whether it achieves its scholarly aims or not, it is my hope that this book might serve to spur on interest in Jonas's thought: both in terms of the eternal problems of philosophy that he attempts to solve and the crucial issues of the present and future that he urges us to address.

Lewis Coyne
Bristol, September 2019

Acknowledgements

Having been so long in gestation this book owes a great deal to a great many people – although its flaws, of course, are my responsibility alone.

My principal academic debts are to the following individuals and organizations: Michael Hauskeller, for his teaching and guidance; the Department of Sociology, Philosophy and Anthropology at the University of Exeter, for providing me with an intellectual home; the Economic and Social Research Council, for their financial support; Roberto Franzini Tibaldeo and Jelson Roberto de Oliveira, for kindly inviting me to develop my ideas at conferences in Louvain-la-Neuve and Curitiba, respectively; Brigitte Parakenings, curator of the Philosophisches Archiv of the Universität Konstanz, for her help in accessing Jonas's papers; my erstwhile peers at the SWDTP in Bristol – Cameron Hunter and Frederick Harry Pitts in particular – for their fine company; and my Exeter-based colleagues, Owen Abbott, Jack Griffiths, Jacob Lucas and Peter Sjöstedt-Hughes, for our discussions in the garden of the Imperial.

This book is derived, in part, from an article published in *Ethics, Policy & Environment* on 12 September 2018, available online: http://tandfonline.com/10.1080/21550085.2018.1509487. In addition, I am grateful to the relevant publishers for granting me permission to reproduce material from the following journal articles and book chapters: Coyne, L. (2017), 'Phenomenology and Teleology: Hans Jonas's Philosophy of Life', *Environmental Values*, 26 (3): 297–315; Coyne, L. (2018), 'An Unfit Future: Moral Enhancement and Technological Harm', in M. Hauskeller and L. Coyne (eds), *Moral Enhancement: Critical Perspectives*, 351–70, Cambridge: Cambridge University Press; Coyne, L. and M. Hauskeller (2019), 'Hans Jonas, Transhumanism, and What it Means to Live a Genuine Human Life', *Revue Philosophique de Louvain*, 117 (2): 291–310; Coyne, L. (2020), 'The Ethics and Ontology of Synthetic Biology: A Neo-Aristotelian Perspective', *NanoEthics*, 14 (1): 43–55. At Bloomsbury, Liza Thompson's encouragement and patience in seeing the project through to completion were greatly appreciated.

On a personal note, I would like to give my heartfelt thanks to my parents and siblings, all of whom have supported my philosophical endeavours, however ill-advised. My hometown friends, too, have with their friendship provided the sort of balance that prevents a book from consuming its author. And finally, my deepest gratitude is, as ever, to my wife, Beccy. Not only did she keep me going throughout the writing of this book, with unending love and care, but she did so while writing a novel herself and giving birth to our little boy, Arthur. This book is dedicated to them both.

Introduction

I Jonas's philosophical project

On 3 September 1939, upon the Allies' declaration of war against the Nazis, the German-born philosopher Hans Jonas wrote the following exhortation to his fellow Jews:

> This is our hour, this is our war. It is the hour we have been waiting for with despair and hope in our hearts [throughout] these deadly years: the hour when it would be allowed to us after powerlessly enduring all the ignominy, all the physical deprivation and moral violation of our people, to finally confront our archenemy eye to eye, with weapon in hand; to demand retribution. (*OPW*: 194)

For Jonas, as the passage indicates, taking up arms against his former compatriots was a matter of honour for Jews, whom the Nazis had above all others marked out for destruction. But he also regarded it as something greater still: a war of two competing spiritual principles. On the one side, 'in the form of Christian-occidental humanity, *Israel's* legacy to the world', on the other, 'the cult of power and contempt for humanity [that] signifies the absolute negation of this legacy' (*OPW*: 199).[1] It was, in short, a world-historical conflict in which Jews had a unique part to play. Jonas lived up to his word, enrolling in the British Army and going on to serve for six years in defence of 'the world in which I as a moral being am able to breathe' (*OFL*: 13).

The thought of this philosopher-soldier, Hans Jonas, is the subject of the present work. Jonas is well known by philosophers in Italy, Israel and Japan, and towards the end of his life was a veritable public figure in his native Germany. And yet Anglophone philosophers, to the extent that they have heard of Jonas at all, are most likely to know him simply as one of 'Heidegger's children' from Richard Wolin's polemical work of that name ((2001) 2015). They may, in addition, be familiar with the unflattering depiction of Jonas in Margarethe von Trotta's award-winning film *Hannah Arendt*, which reduces him to an embittered antagonist of the titular character (2012). To say that these two works do an injustice to Jonas both personally and philosophically is an understatement. He is, in fact, one of the most prophetic thinkers of the twentieth century, and perhaps more than any other philosopher of that time risked life and limb to act on his convictions.

The present work is an attempt to demonstrate the enduring value of Jonas's philosophical system. But what is the nature of that system, exactly? A clue has already been given, in Jonas's declaration of commitment to the marriage of Judeo-Christianity and Greco-Roman antiquity that is 'the [W]estern heritage' (*OFL*: 13). For what the spiritual intertwining of Athens and Jerusalem has bequeathed, through many historical twists and turns, is the 'rational-humane civilization of modern Europe', characterized by an 'ethics of consciousness' and 'respect for man' (*OPW*: 199). The dark irony, however, is that the Western tradition has also given rise to an unprecedented *threat* to human existence, and that of all life on earth, in the form of the ecological crisis. Jonas was one of the first philosophers to recognize this fact, and just as he elected to militarily defend the Western tradition from itself, so too did he attempt to confront the intellectual source of our environmental predicament: the gnosticism of Western modernity.

A full account of what Jonas means by the gnosticism of modernity will be given in Chapter 1. Provisionally, however, we may say that it refers to the technological drive that defines our time. This characterization serves to remind us that the story of the modern West is an ambiguous one. On the one hand, it is a tale of progress, demonstrated above all by the provision of food, shelter, liberty and security to almost all citizens. These are real advances, and should not be taken for granted. On the other hand, however, the story of modernity is also one of loss. Although harder to identify and describe, for Western humanity the world has become *unheimlich* (uncanny, or, more literally, *unhomely*). For us, nature – including human nature – is devoid of cosmic significance and objective value; it is a mundane thing to be manipulated for the sake of greater utility. Thus we have penetrated into the workings of life and have upset the ecological balance more than any civilization before us – all the while unknowingly destabilizing the very grounds of our existence. The consequence is that the modern West, though custodian of the tradition of respect for the person, at the very same time imperils that legacy on two fronts: ecological catastrophe without and biotechnological transformation within.

Such is the ambivalent nature of modernity, which can be neither rejected nor affirmed in its entirety. It must instead be *saved from itself* by establishing moral and metaphysical principles that could set limits on collective action.

But what would be the means of doing so? A renewed faith, perhaps? To this question Jonas answers in the negative. Although he remained a practising Jew, Jonas was also enough of a hard-headed pragmatist to know that religion has lost its wider purchase. For us, God is dead. Nothing is sacred. But does this entail that everything is permitted, as Dostoyevsky's Dmitri Karamazov claims? By no means. Although we are '[a]bandoned to "sovereign becoming" […] after abrogating transcendent being', Jonas suggests that we may instead 'seek the essential in transience itself' (*IR*: 125). And it may well be that the dignity of the person, the 'essential' of which he speaks, indeed has its foundation in this world rather than the next. Proof of this would not come through divine revelation, but instead the recovery of key aspects of the Western tradition: namely, a neo-Aristotelian ontology and a neo-Kantian ethic of responsibility.[2] The attempt to prove the validity of both, with a view to demonstrating our responsibility to the future of human life, is the essence of Jonas's philosophical project, and to do it justice, as I hope to in this work, firstly requires an account of his life and formative influences.

II The man and his work

Hans Jonas was born in 1903 to an upper-middle-class Jewish family in the German city now known as Mönchengladbach. It was, for the most part, a comfortable childhood – so much so that Jonas longed for something dramatic to happen, something comparable to the great battles of antiquity that had captured his imagination. The only hardship the young Jonas endured – aside from the tragedy of his brother's premature death – was the widespread anti-Semitism of the day. When subject to such slurs, Jonas, though slight of stature, would hurl himself at the perpetrator. Even as a child he believed that, belonging to a minority, 'we couldn't take anything lying down; we would never be completely accepted' (*M*: 25–6).

Given this combustible mix of Jewish defiance and longing for purpose, it is unsurprising that the adolescent Jonas turned to political Zionism. At that time the movement to found a state in Palestine was controversial among European Jews, particularly those who enjoyed middle-class lifestyles in societies that appeared to be increasingly liberal. In that context it was not so unreasonable to hope that the intensification of German anti-Semitism after the First World War might simply be a temporary setback, brought about by the humiliation of defeat and subsequent political upheaval. In the Jonas household this debate played out ferociously between Hans and his father, the latter of whom regarded Zionism as both disreputably radical and politically unnecessary, given that normal ethnic relations were sure to resume. Tragically they did not, of course, and with the benefit of hindsight Jonas's preference for emigration from rather than integration into Weimar Germany, though motivated by a young man's fervour, appears prescient.

On finishing school in 1921 Jonas did not immediately head east to Palestine, however, but rather south, to the University of Freiburg. There he crossed paths with a professor's assistant of prodigal talent named Martin Heidegger. Jonas had originally gone to Freiburg to study under Heidegger's mentor, Edmund Husserl, who was by then firmly established as the leading German philosopher of the day. But as a first-year undergraduate Jonas was only permitted to attend Husserl's lectures, not the accompanying seminars, whereas Heidegger's seminars were open to all – provided they could read ancient Greek. If it is remotely fair to reduce a philosopher's intellectual biography to a few key events, then, in addition to Jonas's declaration of commitment to the Western tradition in the face of the Nazi threat, his enrolling in Heidegger's course on Aristotle is one such moment.[3] Aside from a two-year interval in raucous Weimar-era Berlin, Jonas's philosophical education essentially revolved around Heidegger. When the latter took up a professorship in Marburg, Jonas followed. There he formed a deep friendship with Hannah Arendt, and spent four years writing his doctoral thesis on the Gnostic religion under the joint supervision of Heidegger and the theologian Rudolf Bultmann. Heidegger's fusion of phenomenology, existentialism and hermeneutics, his critical reflections on modernity and his radical deconstruction of Western metaphysics were all to exert a profound influence on Jonas's thought.

Needless to say, Heidegger's political alignment with the Nazis in 1933 came as an immense blow to Jonas, both personally and philosophically. Although he claims to have found student hero-worship of Heidegger 'profoundly repellent', there is no doubt that for a time Jonas was also under Heidegger's spell (*M*: 59). Even when this awe had long since diminished, an appearance on the German television programme *Zeugen des Jahrhunderts* in 1982 showed that even after all that time Jonas kept a sketch of Heidegger above his desk.[4] Any lingering fondness would surely have vanished, however, had Jonas known of Heidegger's anti-Semitism, which has only recently been confirmed with the publication of the *Black Notebooks*.[5] As it is, Jonas sincerely believed that Heidegger was no anti-Semite. But whatever Heidegger's motivations, the fact that the 'most profound thinker of our time fell in with the goose-stepping brown shirted battalions' crushed Jonas's faith in the humanizing power of philosophy (*M*: 187).

Although he by no means foresaw what was to come, Jonas wasted little time in emigrating when Hitler came to power. Vowing as he left German soil 'never to return except as a soldier in a conquering army', Jonas set sail for Britain in 1933 and on to what was then the British Mandate of Palestine in 1935 (*M*: 75). It was in Palestine that his academic ambitions were first superseded by martial commitments. Jonas joined Haganah, the paramilitary organization dedicated to defending Jewish settlements, and then in 1940 enlisted in the First Palestine Anti-Aircraft Battery of the British armed forces.

It seems that Jonas found war to be surprisingly conducive to philosophical reflection, as he worked out the second major element of his philosophical system during these years of service. If the first part, his analysis of gnosticism, concerns our spiritual estrangement from the world, the second seeks to overcome it. Far from being alien to nature, humanity is, according to Jonas, the highest manifestation of a certain tendency within it. At the core of this theory is an account of living beings as immanently purposeful, something witnessed in their perpetual struggle for continued existence. Here Heideggerian existentialism collides with a neo-Aristotelian philosophy of nature, the result no doubt owing its starkness to the military context of its development, as Jonas himself acknowledges (*PE*: xiii).

In 1944 the various Jewish units of the British Army were merged into the Jewish Brigade, which was then deployed in the Italian Campaign and the bloody Battle of the Senio. As the Brigade gradually ascended the boot of Italy, word of gas chambers and extermination camps made its way through the ranks. Thus the terrible truth of the Holocaust slowly began to dawn. Returning to Germany in 1945 as part of a conquering army, thereby fulfilling his vow of twelve years previously, Jonas was ashamed to find himself gratified at the sight of great German cities reduced to rubble: the 'evidence of justice, of divine retribution' (*M*: 132). He made his way to his hometown of Mönchengladbach and there learnt of his mother's fate: she had been murdered in Auschwitz in 1942. This was a devastating revelation for which Jonas could not forgive the German people, and he would never again live in his country of birth.

Jonas returned to Palestine and was before long drafted into another conflict, the Arab–Israeli War of 1948. By then in his mid-forties and tired of fighting, Jonas subsequently decided to take an academic post at a satellite college of McGill University in Québec, Canada. Leaving the newly formed state of Israel for the safety of North America opened him to accusations of betrayal by Zionist friends, which seems to have led Jonas to a more nuanced assessment of the cause to which he had long been committed; at any rate, he late in life expressed regret that he and his fellow Zionists had so 'loftily ignored' the interests of Palestinian Arabs (*M*: 37). In 1955 Jonas moved from Canada to the New School for Social Research in New York, famous for its University in Exile constituted by European émigrés. There he was finally at liberty to pursue a conventional academic career.

At the New School Jonas was known to the student body as one of the department's 'Big Three' figures (Rutkoff and Scott 1986: 214). Constituting the final component of his philosophical system, Jonas's thinking there turned towards ethics and political theory, motivated by the unprecedented pace of technological development that occurred in the mid-twentieth century. In the military domain, of course, the great technological acceleration manifested in the form of the atomic bomb and subsequent nuclear arms race. Jonas, however, was less concerned by these obvious dangers than by the 'slow, long-term, cumulative' problems entailed by apparently benign industrial developments (*IR*: ix). In particular, he was quick to recognize that the unfolding ecological crisis and rapid advances in medicine and biotechnology were matters of great philosophical import, and it is through his analyses of both that he went on to enjoy intellectual and political influence late in life.

In his adopted country of the United States, Jonas had his greatest successes in what are now known as medical ethics and bioethics. Although his instincts placed him at odds with mainstream Anglo-Saxon utilitarianism, Jonas nevertheless helped to develop these fields as a founding fellow of the Hastings Center, twice giving evidence to the US Senate in that capacity (*TH*; *THSR*). In his country of birth, by contrast, Jonas became best known for the environmental ethic developed in his 1979 masterpiece *Das Prinzip Verantwortung* (in English translation, *The Imperative of Responsibility*). This work achieved the remarkable feat, unimaginable in the Anglosphere, of quickly selling over 200,000 copies, thereby elevating Jonas to the status of public intellectual. A flurry of accolades and civic activity followed, including addresses on ethical issues of the day, consultation with industrialists and politicians, and receipt of the prestigious Peace Prize of the German Book Trade in 1987.

Long since formally retired, Jonas nevertheless continued to give lectures and interviews into his ninetieth year. His final public word was given in Udine, Italy, on the occasion of his reception of the 1993 Premio Nonino prize for *Maestro of Our Time*. Jonas had forbade himself any further trans-Atlantic travel, yet made an exception to visit Udine. For it was there, almost a half-century earlier, that Jonas the soldier met two elderly Austrian Jews who, having fled the Nazis, found themselves protected and provided for by the ordinary townsfolk. Jonas says that he kept this memory within him 'like a sacred trust', and one suspects that its importance lay in helping to keep his faith in humanity alive just as he was learning of the Holocaust (*MM*: 200). Having recounted the story in Udine, Jonas returned home to New Rochelle, New York, and died six days later. A statue of him now stands in the Hans-Jonas-Park of Mönchengladbach.

In reconstructing, analysing and occasionally amending Jonas's philosophical system, the present work will proceed as follows. We shall begin with the most immediate issue, namely, his critique of the gnosticism of modernity, which has both theoretical and practical sides. The former concerns the gnostic world view, which, in both its ancient and modern manifestations, construes humanity as alien to the natural world. The practical side of Jonas's critique concerns three quandaries the modern West finds itself presented with as a result: nihilism, ecological ruin and the possibility of our own biotechnological transformation.

Since his critique locates the source of gnostic error in its ontological commitments, Jonas's attempt to overcome it begins with a new philosophy of nature.[6] He argues that connecting all life, from the lowliest unicellular organism through to plants and animals and human beings, is a *freedom* unknown – except in latent form – to inanimate matter. Since human beings represent the pinnacle of life's movement towards world-openness, Jonas's philosophy of biology leads, as a matter of course, to a philosophical anthropology. In this way Jonas seeks to overcome the gnostic vision of a humanity estranged from the natural world, at the same time as accounting for our unique place within it. Holding that the nihilism of our time follows directly from modern gnosticism, Jonas then seeks to prove that an ethics can be derived from his alternative ontology. As such, we shall turn to axiology and then moral philosophy. Jonas argues, in short, that all living beings are ends-in-themselves courtesy of their capacity to value, and it is this that grounds our responsibilities towards them. Human beings alone can *have* responsibilities, however, and in this ethical capacity lies our unique personal dignity. The 'idea of Man' as a moral being, as Jonas puts it, is accordingly the ultimate object of our responsibility, and the imperative to safeguard its realization in human beings an unconditional obligation.

Our final chapters deal with the political philosophy and practical ethics Jonas develops to counter modernity's most pressing concerns. Because the ecological crisis is an inherently collective rather than individual phenomenon, driven by a gnostic utopianism, the solution must be an equally far-reaching counter-utopia. On this basis Jonas attempts to devise principles, based on our responsibility for the idea of Man, that might underwrite new environmental policies and institutions. Lastly, we shall look to Jonas's bioethical contributions, in particular his analysis of genetic engineering as a means to the end of human enhancement. Rejecting the utopian drive to 'perfect' humanity – a spurious notion if ever there was one – Jonas bids us to newly appreciate 'genuine man', who 'is always already there and was there throughout known history: in his heights and his depths, his greatness and wretchedness, his bliss and torment, his justice and guilt – in short, in all the ambiguity that is inseparable from his humanity' (*IR*: 200).

Such are the key components of Hans Jonas's philosophy.[7] At this point, even his most sympathetic interpreter would have to admit that although Jonas accurately describes our plight, his efforts to find the solutions do not always succeed, and occasionally stand in need of corrections of their own. When departing significantly from Jonas's own thought – as I believe we must in the cases of his axiological objectivism and critique of liberal democracy – I will explain where he errs in his attempt to overcome the gnosticism of modernity. The attempt itself, however, is always commendable. Remaining true to the combined legacy of Athens and Jerusalem, the West's 'eternal origins', Jonas's philosophy both alerts us to our present plight and goes some way towards showing how we might overcome it (*OR*: 3).

1

The gnosticism of modernity

I The gnostic principle

Jonas's philosophical system, from his metaphysics and his ethics to his political theory, is a corrective to modern gnosticism and its threefold legacy: nihilism, ecological ruin and biotechnological transformation. But what exactly is gnosticism, and what is peculiar to its modern form? We said in the Introduction that the latter could be provisionally understood as another name for the technological drive that characterizes our time, but this was only a cursory definition. The purpose of the present chapter, therefore, is to properly define gnosticism and explain the significance Jonas attributes to it. Although, as indicated, modern gnosticism is Jonas's ultimate object of concern, to understand the latter we must begin with an account of its original incarnation.

The Greek word *gnōsis*, from which the term 'gnosticism' derives, simply means knowledge, and used in this everyday sense it can be found throughout the works of the great Greek thinkers. Gnostic*ism*, however, refers to something more specific: namely, a world view with certain key themes and tropes.

The history of gnosticism begins in the period of antiquity defined by the Roman Empire, dated from the latter's foundation in the first century BCE through to fall of the Western Empire in 476 CE. Over the course of this half-millennium the Abrahamic religions, in many ways at odds with the old Greco-Roman paganism, gained prominence across the Mediterranean world. The Roman emperor Constantine's conversion to Christianity in c.312 CE was a symbolic turning point in this process, as was the legalization of Christianity throughout the empire in the following year. Meanwhile, beyond the walls of Rome various ascetic cults and sects had taken root. Christian monasticism, which would later become a mainstream form of that religion, developed in Egypt; in Syria, Stylites demonstrated their devotion to God by praying and fasting atop pillars. Perhaps most bizarrely, the Hermetics developed a cosmology that fused

Mediterranean paganism with monotheism, alchemy and magical practices. In terms of independence of character and philosophical depth, however, the most significant such sect were the Gnostics.

Emerging in the first and second centuries CE, the Gnostics were, according to Jonas, something of an oddity even in their own time. Although their thought owed most to Christian scripture, it was nevertheless a 'wild offshoot' from conventional Christianity, informed by Platonic, Judaic, Persian and Egyptian elements (*WPE*: 30). Besides being of interest in its own right, Jonas holds that the Gnostic religion was also of philosophical significance, and to this end his interpretation seeks to uncover the underlying and unifying character of the Gnostic texts. Using Heidegger's existentialism as an interpretative framework, Jonas's analysis identifies the essence of Gnostic thought as its 'radical and uncompromising' understanding of being, both in terms of individual existence and the nature of the cosmos as such (*GR*: 26).[1]

The radicalism of the Gnostic religion is most clearly illustrated by comparing its basic cosmology with that of Christianity. The latter is summarized by Jonas as follows:

> The created world of Genesis is not a god and is not to be worshipped instead of god. [...] Jewish monotheism had abolished the deities of nature and all intermediary powers, leaving God and world in clean-cut division. The Christian hierarchy of angels and saints did not bridge the gulf between God and the world, but that between God and the human soul which, not being of the natural order itself, shares their preternatural status. (*PL*: 71)

This notion of God and soul being at odds with the natural world also formed the basis of Gnostic cosmology, although in more extreme form. According to the Gnostics humans possess a divine spark, a vestige of God, in the form of a soul. Clearly, this much is in line with Christian doctrine. Yet the Gnostics went considerably further in their cosmological dualism, as they held the physical world to have been created not by God, but rather by the 'demiurge': a demonic entity, sometimes depicted as a snake-like chimaera, which imbued the natural world with darkness. As such, the God of the Gnostics is wholly removed from the world:

> The cardinal feature of gnostic thought is the radical dualism that governs the relation of God and world, and correspondingly that of man and world. The deity is absolutely transmundane, its nature alien to that of the universe, which it neither created nor governs and to which it is the complete antithesis: to the divine realm of light, self-contained and remote, the cosmos is opposed as the realm of darkness. (*GR*: 42)

This stark dualism extended to the Gnostics' understanding of human beings themselves. According to the Gnostics, each human soul had been ensnared by the demiurge before being taken from the world of light to the material domain, and there imprisoned in a body. As such, in the Gnostic religion the human essence is wholly alien to the physical world it inhabits: each divine soul is shackled to a conscious mind, which is in turn incarcerated in a body, and stranded on an unholy world (*GR*: 44).

The Gnostics held, nevertheless, that each individual has a chance of escaping its fallen condition in death. There the soul may find a release from the material plane provided one prepares in life by arriving at a true understanding of the world. Study of the Gnostic texts – made possible by a 'messenger from the world of light', most commonly identified as Jesus – reveals the truth of the natural world and the demiurge, and so allows the enlightened soul to find its way back to the world of light (*GR*: 45). This is the revealed knowledge, the *gnōsis*, that gives the religion its name, and adherence to which represents the promise of salvation. In addition to tempering an otherwise strikingly bleak world view, this eschatological dimension rescues Gnostic thought from absolute nihilism: an objective good is believed to exist, albeit aligned with the unknown God and hidden from us by the dark shroud of our given reality.

For the purposes of this work the preceding account of Gnostic cosmology, though brief, will suffice – interested readers should consult Jonas's masterwork of scholarship, *The Gnostic Religion*, which analyses the ancient texts in detail. What concerns us here is the *significance* he attributes to the religion. Its appearance is, to be sure, a curious episode in Western history. But is it any more than that? Jonas argues in the affirmative, for two reasons. The significance of the Gnostic religion lies firstly in that it tells us something about the period of antiquity in which it took root and grew; secondly, and more importantly, it sheds light on Western history as such. For it is Jonas's contention that the essence of the Gnostic world view – which we shall call gnostic*ism* or, following Jonas, the 'gnostic principle' – recurs throughout Western history, informing even our own time (*GR*: xxxv). These are bold claims, and so we shall unpack and attempt to justify them in turn.

The Gnostics held, as we have seen, that the world in which we live is fundamentally opposed to the good, the latter belonging to the alien God and world of light alone. In this cosmic pessimism it constituted, according to Jonas, the pinnacle of nihilism in its period of antiquity. Although this does not make Gnostic thought '*the* key to understanding the whole epoch' – which is generally less apocalyptic – it nevertheless illuminates the time from which it emerged (*GR*: 26). Most notably, the Gnostics' identification of this world with malignant darkness represents a radical break with the culture of Classical antiquity – a break so severe that it was only credible in a historical moment of deep transformation.

Traditionally, Greco-Roman thought had held that the public realm of the *polis*, or *res publica*, made the good life possible, as only by living in such a polity could humanity fully realize itself in theoretical enquiry and political action. This world of human affairs, watched over by the gods, was thought in turn to be situated within the eternal workings of nature, together forming a harmonious and ordered cosmos (*PL*: 222). Thus to the Classical mind what-is (Being) was intimately connected to what-ought-to-be (the good), a holistic perspective that formed the background for life in the Greek city states and Roman Republic of the Hellenic and Hellenistic eras.[2] However, following the transformation of the Roman Republic into the Roman Empire, active citizenship in the public realm was no longer possible – hence the good life, as it had been conceived of until then, was no longer attainable. Jonas suggests that the Roman intelligentsia attempted to compensate for this loss by adopting the cosmopolitan philosophy of Stoicism, but the shift away from Classical Greco-Roman thought nevertheless transformed the long-established connection between Being and the good life (*PL*: 223). While this rupture gave rise to various eccentric philosophies and cults, none articulated this crisis more dramatically than the Gnostics, for whom, as we have seen, the good was conceived of as wholly separate from – and indeed, unrealizable within – the material world.

Such is the meaning Jonas attributes to the religion with regard to its own time: it is the most radical philosophy to spring from the demise of the Greco-Roman public realm. As stated, however, he also claims that the *essence* of the religion transcends its original manifestation. This essence consists of the following three principles, distilled from the religious Gnostic texts:

1) The soul is separate from nature.
2) Nature has no value save that of providing for the soul.
3) The soul must transcend nature in death.

In various manifestations these notions have, Jonas suggests, recurred throughout Western civilization, to varying degrees informing the zeitgeist. At certain times and places, such as the Florentine Republic, the gnostic principle partly retreated from view, while at others – most obviously, perhaps, the High Middle Ages – it came to dominate an entire epoch. Jonas contends, however, that gnosticism finds its most extreme and consequential manifestation in the *modern* world. At first glance this claim appears improbable: modernity is secular and rational, with a technologically advanced capitalist mode of production, all of which could not be more different to the Gnostics' own period of history. Jonas suggests, however, that on a deeper level the gnostic principle profoundly informs the spirit of our time.

It should be noted in advance that although Jonas's analysis of gnosticism will prove to be overwhelmingly critical, he does attribute one significant positive effect to the phenomenon. By stressing the dissimilarities rather than the commonalities holding between humanity and nature, and by subsequently devaluing the latter, gnosticism turns humanity in on itself in search of meaning. Although meditative cultivation of the self pre-existed gnosticism, as evidenced by the Orphic mysteries, it is nevertheless 'to the long reign of dualism [...] that we in the West owe the exploration of the realm of the soul, indeed the enrichment of the soul through constant reflection' (*MM*: 47). The psychological depths plumbed by writers from St Augustine through to Kierkegaard and Dostoyevsky can be attributed to precisely this tendency; the same is true, of course, of the vast inner landscape revealed by the dualistic religions and philosophies of Asia. Since this cultivation represents a real gain we are all, to a degree, beneficiaries of the gnostic principle. Alas, the price paid for this enrichment of mind was high, in modernity taking the form of not only a nihilistic estrangement from the world but also the present environmental crisis and the prospect of our own biotechnological transformation. The remainder of this chapter will seek to justify this audacious claim.

II Nihilism, ancient and modern

Jonas's discovery of the spiritual parallels between modernity and the Gnostics' period of antiquity is the result of a curiously dialectical process, one that requires further biographical information. As indicated, Jonas uses Heidegger's existentialist concepts to give voice to the Gnostic world view.[3] It might seem, perhaps, to be an odd choice of interpretative framework, yet the parallels are striking in several respects. In order to sketch out this connection, we shall give a brief account of Heidegger's thought as it had developed when Jonas was studying under him.

Heidegger's lifelong concern was the 'question of the meaning of being', an issue he would eventually approach from a great many angles (2010a: 4). In his early work, however, which culminated in the 'revolutionary' *Being and*

Time, Heidegger sought to clarify that question by asking after the being of the entity that could pose it (*HBT*: 1).⁴ That entity was *Dasein*. The word *Dasein* is a common enough German noun typically translated into English as 'existence', although a direct translation of its component words would be 'being-there'. Heidegger alluded to both senses, using the term to denote those beings that are concerned with their own being, and it is clear that when doing so he primarily had human beings in mind.

The most significant feature of *Dasein*, according to Heidegger, is the fact that it 'is concerned *about* its very being', and thus, however unconsciously, 'understands itself in its being' (2010a: 11). By this Heidegger meant that in its conduct and practices *Dasein* always demonstrates an implicit understanding of its possible ways of being: taking up one project and not the other, for example, or acting in one way rather than another. This understanding of the possibilities of its own being Heidegger called *Dasein*'s 'existence', alluding to the latter term's etymological meaning of standing out or standing forth (2010a: 11). To make sense of this notion consider that human beings never *abide* in the way that a book, a mountain or a hammer does, say. On the contrary, we are always one step ahead of ourselves in being oriented towards possible tasks to be carried out and goals to be accomplished; simultaneously, we are always one step behind ourselves, as it were, in that a history (both personal and cultural) informs those projects. Facing ahead of itself, based on its past, the being of *Dasein* is *temporally* structured – hence the title of Heidegger's great work.

Since *Dasein*'s (for the most part implicit) understanding of its own possibilities constitutes its existence, Heidegger sought to establish from where this understanding originates. The answer is the everyday being of the world into which *Dasein* is 'thrown' (2010a: 131). By the latter Heidegger did not mean that we were *literally* thrown into the world by a higher power, as the Gnostics had thought, but rather that *Dasein* finds itself in the world. More specifically, it finds itself in a particular social world that provides but a partial selection of the possible ways in which *Dasein* can be. Thus we typically act as *they* act, think as *they* think, speak as *they* speak. Here 'the they' (2010a: 124) does not refer to a specific group of people, but rather the abstract social subject invoked in phrases such as 'what one does'. Whether in conformity with or in rebellion against it, the social world in which *Dasein* finds itself provides the basis of its understanding of its possibilities.

Heidegger's existentialism manifested most vividly, however, with the significance he attributed to our mortality. According to Heidegger, *Dasein*'s projections into the future are bounded by the knowledge that it must one day die. As the negation of all possible ways of being, death represents 'the possibility of the absolute impossibility' of *Dasein* (2010a: 241). Heidegger called living with this certainty our 'being-toward-death', and he thought that we typically flee from an honest confrontation with it by taking refuge in the possibilities made available to us by the they (2010a: 241). In other words, Heidegger thought that we flee from our being-toward-death by identifying with the concerns and customs of the society we have been thrown into, pretending that these are essential to our being. The inauthenticity of such an existence is revealed to us in moments of existential anxiety, for we then see that in comparison to the fact that we must die – our one 'insuperable possibility' – all other possibilities are merely arbitrary (2010a: 241). In anxiety, as Heidegger put it, *Dasein* is fetched 'back out of its entangled absorption in the "world". Everyday familiarity collapses. *Dasein* is individualized', and in the process brought before 'the authenticity of its being' as existence (2010a: 182).

What, then, does Jonas see of existentialism in Gnostic thinking? The connection does not pertain to Heidegger's broader notion of existence – which, as we shall see in Chapter 2, Jonas himself draws on – but rather Heidegger's analysis of *Dasein*'s everyday understanding of itself. The first critical point of comparison is the notion of being 'thrown'. As we have seen, the Gnostics held that humans do not belong to the malign natural world, but were instead cast into it by the demiurge. Jonas suggests that this religious notion finds a secular equivalent in Heidegger's claim that *Dasein* finds itself in being without explanation (*PL*: 233). As Jonas points out, the latter can only make sense if nature is understood not as a creative force – or the cradle of civilization, as the Greeks had thought – but rather as 'neutralized' and 'indifferent', a 'bare nature' reduced 'to the mode of mute thinghood' (*PL*: 231). For Jonas, therefore, thrownness is not a universal and timeless existential structure of *Dasein*, but rather a historically contingent one.

In a further parallel, both Heidegger and the Gnostics held that most of us, most of the time, avert our eyes from (what they deemed to be) the truth of our existence. The Gnostics believed that we are led astray from salvation by the distractions of the world; for Heidegger, the temptation to flee from our being-toward-death similarly lay in inauthentically succumbing to the concerns of mass humanity. For both, therefore, it is the very public world held by the Greeks and Romans to be our place of greatest self-realization that in fact *prevents* us from seeing the light. Moreover, both Heidegger and the Gnostics considered these worldly temptations to be ones that the individual can, in principle, resist. For the Gnostics, all humans – with the exception of 'somatic' individuals who are irredeemably devoted to bodily pleasures – can acquire *gnōsis* through study of the relevant teachings. Thus the soul, turning away from the world, prepares itself for salvation, while for Heidegger, as we have seen, each of us as *Dasein* has the potential to gain an authentic grasp of our being. As such, according to both, humanity can either grasp the true circumstances of its being or else succumb to the 'tranquilizing alienation' of worldly affairs (2010a: 171).

On the basis of such affinities Jonas is able to use Heidegger's existentialism to give voice to Gnostic thinking. But this very exercise also leads him to ask *why* the former is such an appropriate framework for understanding the latter, which would seem, after all, to be quite a remarkable coincidence given the millennia separating them. To this end Jonas inverts his interpretative enterprise, and uses the Gnostic tenets to illuminate existentialism. 'In other words, the hermeneutic functions become reversed and reciprocal – lock turns into key, and key into lock: the "existentialist" reading of Gnosticism [...] invites as its natural complement the trial of a "gnostic" reading of Existentialism' (*PL*: 213). The effort leads Jonas to conclude that their underlying affinity, the reason for their interpretative fit, lies in a shared nihilism. For Heidegger, as we have seen, *Dasein* typically fails to attain authenticity by unthinkingly partaking in the pattern of life offered by one's culture, a state he called fallenness. For the Gnostics, meanwhile, we are trapped on a material world conceived of and created in darkness by the unholy demiurge. Although both are obviously despondent, the vivid language used by the Gnostics to express their world view might give the impression that it is more nihilistic than Heidegger's. This would be a mistake, however. Jonas's comparison of Gnostic and existentialist nihilism leads him to the very opposite conclusion, namely, that the latter is far more extreme than the former – an insight that is pregnant with significance.

Jonas's reasoning is as follows. Although the Gnostics took the public and natural worlds to be debased, they nevertheless retained the belief in a divine realm to which the soul could return in death; thus the Christian dualism of heaven and earth was upheld. For existentialism, however, there is no escape from the purposelessness of our existence. There is only death. According to Heidegger, the best one can do is face the arbitrariness of our possibilities, revealed to us in moments of existential anxiety, with 'resoluteness' (2010a: 323). The consequence is this: although the Gnostics held that we inhabit a malign world, their vision was still one of *objective* good and evil, thus preventing a slide into absolute nihilism. For the existentialist, however, there are no such objective values, good or evil. The world is instead devoid of *any* given meaning and lacks *any* fixed norms by which we might individually and collectively live.

Accordingly it is this world view, as Jonas says, that constitutes 'the true abyss. That only man cares, in his finitude facing nothing but death, [...] is a truly unprecedented situation' (*PL*: 233).

In spite of their affinities, then, the absolute nature of existentialism's nihilism marks it out as more radical even than the Gnostic religion. For Jonas the question then is what wider historical and cultural development allowed for it, explaining such a difference. In search of an answer he firstly looks again to the Gnostics in relation to their own time.

The Gnostic religion was made possible, we noted, by the collapse of Classical Greco-Roman civilization, which had held the good to be realized within the public world of the *polis*, in turn situated within the eternal workings of nature. The Gnostics, by contrast, believed in an objective good divorced from both nature and society, and defined worldly affairs by deceit alone. Here we have, according to Jonas, the core of the religion's nihilism: *its denial of any connection between the good and the given world.*

With this insight Jonas's analysis appears, initially at least, to align with Nietzsche's great diagnosis of nihilism – a comparison that is worth drawing out. Nietzsche famously held that any attempt to justify existence through transcendent principles risked succumbing to nihilism through a devaluation of nature. His chief culprits in this regard were Plato, who held the material world to be a mere copy of the world of Forms, and Judeo-Christianity – which he pithily dubbed 'Platonism for "the people"' – for its belief in a benevolent creator deity (1990: 32). From Jonas's perspective Nietzsche's analysis is largely correct, as the gnostic principle has typically appeared throughout Western history under the guise of a transcendent explanation for existence that divorces humanity from nature and devalues the latter.[5] But modern nihilism – exemplified, for Jonas, by Heidegger's existentialism – is unique in its depth, and yet makes no appeal to transcendent principles. On the contrary, it holds that there is *only* humanity and the meaningless world. As such, the deepest manifestation of nihilism fails to fit the cause suggested by Nietzsche. But why was this?

Jonas argues that Nietzsche failed to identify the true cause of modern nihilism because this cause is, in fact, a metaphysical belief that Nietzsche himself shared. In his posthumously published notebooks, Nietzsche tells us that 'the general character of existence does not admit of interpretation in terms of the notions "*purpose*", "*unity*" or "*truth*". Existence achieves nothing and accomplishes nothing' (2017: 19). Jonas notes that this is a thoroughly *modern* view, unknown to Plato, Christians and Gnostics alike. Though all were nihilists to varying degrees, courtesy of their relative denigration of the physical world, they at least afforded nature some significance by acknowledging its place in an ordered cosmos. The *truly* 'disenchanted world', however, 'is a purposeless world', and for all their devaluation of nature Platonism, Christianity and gnosticism did not construe it as such (*PE*: 171). Modernity alone denies purpose, order and objective value to nature in its entirety, and it is *this* belief that explains the nihilism of our time – a development far more thoroughgoing than the collapse of Classical Greco-Roman civilization that had led to the Gnostics' nihilism. As Jonas puts it: '[A] change in the vision of nature, […] of the cosmic environment of man, is at the bottom of that metaphysical situation which has given rise to modern existentialism and to its nihilistic implications' (*PL*: 216).

Before elaborating on what this changed vision of nature consists in, let us recap the argument so far. Nihilism in general, according to Jonas, follows from a severing of the link between the good and the world in which we live. The peak of ancient nihilism was the Gnostic religion, which held the material world to be objectively evil, a belief made possible as a cultural phenomenon by the loss of the *polis* and *res publica*. Modern nihilism, exemplified by Heideggerian existentialism, goes yet further, however. The cause of its unique depth is the modern belief that the world is alien to *any* purpose, order or objective value – positive or otherwise. But how did this come to pass? How did the West move from the belief in an ordered cosmos, which in very different forms nevertheless underpinned both antiquity and the Christian Middle Ages, to the purposelessness of modernity?

Jonas argues that it can be traced to the metaphysical revolution marking the birth of the modern West. This revolution consisted of a turn away from the old Aristotelian-Scholastic understanding of Being and towards materialism, paving the way for not only modern nihilism but also the ecological crisis and the development of biotechnology. It is, therefore, a historical event of the first order, but one little understood. The significance Jonas attributes to this metaphysical revolution constitutes the core of his critique of modernity, and is the principal issue to which his entire philosophical system is a response. Before explaining it in detail, however, the key points will be summarized.

We noted earlier that the three core features of gnosticism were, first, a dualistic metaphysics separating humanity from nature, second, the relative devaluation of the natural world and, third, a utopian eschatology that promises (at least part of) humanity an escape from its fallen condition. In the Gnostic religion, Platonism and Christianity, these features appeared under a transcendent guise. The Gnostics, for instance, conceived of Being as split into the holy world of light, to which the alien God and human souls belong, and the malign material world created by the demiurge, to which enlightened souls could escape in death. In modernity's metaphysical revolution, however, the three core features of gnosticism are *immanentized*, meaning that they characterize a vision of Being that eschews any claim to transcendence. More specifically, in modernity gnosticism manifests in the following three motifs:

1. Mind is separate from matter.
2. Matter has value only insofar as it provides for the mind.
3. Mind must use technology to transform the material world.

Through its immanentization the gnostic principle is reborn as pertaining to this world alone. We shall now look at this development in detail, beginning with how modernity's metaphysical revolution gradually led to the absolute nihilism represented by existentialism.

III The scientific revolution

Jonas first explored the significance of modern metaphysics' anti-Aristotelian turn in the 1959 essay 'The Practical Uses of Theory' (*PL*: 190), but his thoughts on the matter are more fully developed in 'The Seventeenth Century and After: The Meaning of the Scientific and Technological Revolutions'. Jonas opens the latter essay by stating, in characteristically dramatic fashion, that '[w]e live in a revolution – we of the West – and have been living in one for several centuries' (*PE*: 46). This revolution is no short-term political event, like the French or Russian revolutions, but rather a 'revolution of thought' that has unfolded over the last five hundred years up to the present day (*PE*: 48).

Despite unfurling gradually, we may point to a single year as symbolizing the beginning of this 'change in theory, in world-view, in metaphysical outlook' (*PE*: 48). In 1543 both Vesalius's *On the Fabric of the Human Body* and Copernicus's *On the Revolutions of the Celestial Orbs* were published. Jonas contends that together they pointed towards the emerging reconfiguration of nature as uniform matter: the former at the 'microcosmic' level of the human body, and the latter at the 'macrocosmic' level of the solar system (*PE*: 52). The materialist interpretation of the human body would eventually result in the loss of a normative conception of human nature – with profound consequences, as we shall see when turning to the question of biotechnology in Chapter 7. The more immediately consequential event, however, was the Copernican mathematical proof of heliocentrism, and so we shall focus on it here.

The reason Jonas takes Copernicus's proof to be pivotal for modernity is that it allowed a plausible materialist understanding of non-human nature to take hold in the modern mind. Of course, the concept of matter had been around since antiquity in various guises: as 'substance' it was at the core of Aristotle's understanding of Being, which then fed into medieval Scholasticism. The difference is that in Aristotelian and Scholastic thought matter was also taken to be teleologically significant in two ways: at the level of individual beings and at that of Being as a whole. To the extent that these metaphysical principles survived the scientific revolution they remain somewhat familiar to us, while striking us in other respects as wholly unfamiliar, and thereby deserve to be spelt out.

First of all, Aristotle held that both natural and artificial beings are characterized by purposes. According to his general metaphysics, the existence of a thing could be attributed to four causes: the formal, material, efficient and final. Take a ceramic vase, for example. The formal cause of the vase is its distinctive shape, the material cause is the clay it is composed of and the efficient cause is the action of the potter who threw it on a wheel. All contribute to making the vase what it is. The final cause is the *telos*, 'for the sake of which a thing is done' (1984g: 194b). When we talk about something's purpose, end or goal, it is precisely its *telos* that we have in mind. We can say, for example, that the vase is teleological in that it exists to hold water: this is its *telos*. There are, however, two different kinds of *teloi*, which we may call *immanent* and *transcendent*. The latter is the sort of *telos* that characterizes a machine, tool or other crafted object, such as the vase. As the term is used in this context 'transcendent' does not necessarily refer to a divine entity or higher power, but simply means that the origin of the *telos* in question is separate from the being that instantiates it. In the case of the vase, we can distinguish between the vase having a *telos* and the origin of that *telos* – the potter – which evidently transcends the vase itself. Understood in this sense, Aristotle's teleology remains readily intelligible to us.

An immanent *telos* differs from a transcendent *telos* in terms of its origin: it both belongs to the thing and originates in it – hence 'immanent'. For Aristotle this kind of *telos* principally characterizes the growth, development, movement and behaviour of living beings. In other words, the *teloi* of living beings lie in their activities and self-constitution, such that they are permanently at work in continuing to be (*entelecheia*) (1984b: 1050a). As natural entities, moreover, their purposes or ends have not been imparted from without, but instead belong to the organisms themselves; the originating and instantiating being of each *telos* is one and the same. This principle may strike us as odd, but we shall see in Chapter 2 that we still have good reason to believe it to be true of all life.

The second sense in which Aristotelian metaphysics regarded matter as teleologically significant was through the gradated structure of Being itself, sometimes called the *scala naturae* or 'great chain of Being'. For Aristotle, nature presents itself to us as hierarchically ordered, from non-living matter, to vegetative life, on through to animals and with humanity atop them all – a gradation that may be explained by each type of being's soul. While this might sound odd to modern ears, for the Greeks the soul was not an exclusively human source of personal identity, but rather an animating force that manifested as the final causes of all living beings. Though the soul was absent from non-living beings, all life was characterized by at least one of its five parts: 'the nutritive, the appetitive, the sensory, the locomotive and the power of thinking' (1984e: 414a). Plants, according to Aristotle, have only the first kind, which pertains to nourishment and growth. Animals have, in addition to this capacity, those of feeling, perception and locomotion, while humans have all four plus the power of thought. Thus nature, viewed in terms of the capacities to which it gave rise, constitutes an ascending scale.

While both of these ontological principles underwent transformation during the Christian Middle Ages – the gnostic spirit of which emphasized transcendent metaphysical principles – they nevertheless survived in recognizable form in the medieval mind, principally through St Thomas Aquinas's reading of Aristotle. With the Copernican revolution and the toppling of the old geocentric perspective, however, the notion of a hierarchically structured cosmos took a first, crippling blow. The critical point is not that the sun was subsequently thought to be the centre of the solar system instead, which is of course factually correct. It is rather, as R. G. Collingwood notes, that the mathematically interpreted universe lacks *any* centre, instead consisting simply of a grid populated by bodies moving between various points (1945: 97). The importance of the proof of heliocentrism, then, was not really astronomical but rather *metaphysical*, entailing the loss of the earth's exceptional status in the universe. Thus began the gradual development of a new understanding of nature as 'homogenous' by virtue of the fact that the earth 'had become a "star" itself [...] and by the same token the planets had become "earths"' (*PE*: 53).

Following Copernicus a host of great figures – Bruno, Galileo, Kepler, Newton – made discoveries in physics and astronomy that cemented and built on his findings. Although as individuals they frequently held nuanced metaphysical positions, their shared success in mathematically interpreting nature lent credence to the metaphysics of uniform matter. Hence it was possible for Galileo to state, in a way that would have been incomprehensible only a few centuries previously, that the 'all-encompassing book' of nature was 'written in a mathematical language' (2008: 183).

Reconciling this new metaphysics with immediate experience was straightforward enough with regard to planets and other non-living beings, which do appear to move without purpose or intent. The puzzle, however, was how living beings – in particular humans and animals – fitted into this new world view, given that they seem to act in precisely the purposeful fashion Aristotle described. The solution to this problem lay in the immanentization of the first key feature of gnosticism: its metaphysical dualism. Under the influence of modern science and mathematics, of which he was an accomplished practitioner, Descartes devised a new metaphysics known as substance dualism. Infamously dividing Being into *res extensa* and *res cogitans*, or 'extended' and 'intelligent' substances, Cartesian substance dualism tore the natural world asunder (1968: 57). The former substance, *res extensa*, was the world of matter as described by the new physical sciences, which, as indicated, could be sufficiently accounted for by material, formal and efficient causation: *contra* Aristotle, no recourse to teleology was deemed necessary. The latter substance, *res cogitans*, was the mind, which accordingly became the sole natural possessor of *teloi*.

Undoubtedly the most radical aspect of Cartesian substance dualism was its extension of materialism into the realm of living beings. In Descartes's mechanistic biology non-human life was stripped of teleological causation: to explain the workings of plants and animals 'it is not necessary', he claimed, 'to conceive of any vegetative or sensitive soul or any other principle of movement and life' (1972: 113). On the contrary, their activity could instead be understood as simply 'educed from the power of matter' (1968: 76). According to Cartesian substance dualism only humans, possessing minds, could act with purpose. As is well known, however, this claim led to a conundrum in that the mind's interaction with the material body could not be accounted for. Descartes recognized, to his credit, that their relation was not simply like that of 'a pilot in his ship', but rather that the mind was 'joined and united more closely with the body' (1968: 76). However, his notorious solution – that the point of connection was the pineal gland in the brain (1972: 95) – was obviously no solution at all. For how, then, was the mind joined to the pineal gland? No adequate answer conceived of on dualistic terms was forthcoming.

Regardless, Descartes's substance dualism set the broad parameters for ontological enquiry over the next two centuries. The effort to resolve its fundamental flaw – that it could not account for the intertwining of mind and matter in humanity – would entail a move to substance monism. As such, two competing positions arose, each emphasizing one half of the substance duality over the other: the idealist path sought to define the material on the terms of the mental; the second option, a pure materialism, hoped to reduce the mental to the material. By the end of the nineteenth century the latter had won out. Darwin's theory of evolution by natural selection proved decisive in resolving the stand-off in favour of materialism, as by extending his insights the materialists argued that humanity, and consequently its mental faculties, were also mere outputs 'thrown up in the mechanics of organic mutation', as Jonas puts it (*PL*: 127). The explanatory advantage of seemingly accounting for the workings of nature thereby allowed a purified materialism to emerge as the dominant ontology of the modern era. As Jacques Monod would eventually claim: '[A]nything can be reduced to simple, obvious mechanical interactions. The cell is a machine. The animal is a machine. Man is a machine' (quoted in Lewis 1974: ix).

With this final development the last remnants of Aristotelian ontology were swept away by the materialist tide. What began with the Copernican revolution as a scientific explanation of the movements of heavenly bodies became, a few centuries later, a totalizing metaphysics encompassing not only non-living beings, plants and animals but human beings also. Belief in the mind's irreducibility to mechanistic nature, which had preserved our dignity, accordingly became implausible, and as a result humanity ceased to believe in its own uniqueness and exceptional position in the universe. As Monod said, all nature was merely matter in motion, uniform, purposeless and subject to causal laws.

Almost as a matter of course, modernity's immanentization of the first principle of gnosticism brought the second in its wake. While there is no *necessary* connection between metaphysical dualism and the devaluation of nature, they are clearly closely connected, and in the history of the West, at least, have largely coincided. For the Gnostics themselves, as we have seen, good and evil were starkly divided between the world of light and that of nature, following from the purported metaphysical dualism that held between the two. And something similar, yet more radical, is also true of modernity.

With the Cartesian restriction of teleology to mind, and mind to humans, purposes were stripped from non-human nature. The entire domain of valuation then became the sole preserve of human beings. Why? Firstly, because once plants and animals are denied purposes nothing matters *to them*. As merely blind mechanisms, unaffected one way or the other by whatever fate befalls them, it makes no sense to speak of their welfare, interests or subjective good, all of which humanity alone is thought to enjoy. *We* may find value in nature, therefore, but if we do – and we could just as well not – this is purely a human inscription on the blank slate of nature. Secondly, with the demise of the idea of a hierarchical and ordered cosmos – the *scala naturae* – the notion that nature as such had ends lost its foundation. To suggest that life in general and humanity in particular represented cosmic achievements, the pinnacle of a creative tendency in Being, became nothing more than speculation. We are simply one more product of evolutionary chance, no more or less significant than the next. As Nietzsche had said, with brilliant finality, existence achieves nothing and accomplishes nothing. With this development the stage was set for the absolute nihilism marking out modernity to triumph.

Such is Jonas's account of our present-day spiritual condition. But can modern nihilism be defeated, and if so, how? Jonas believes that it can. To do so we would have to tackle it at its roots: overcoming the metaphysical revolution that progressively stripped the dignity of purposes from non-human nature and demolished the *scala naturae* (PL: 284). We would, in other words, have to disprove Nietzsche's claim that 'the categories "purpose", "unity", "being" are no longer available to us'. Nietzsche took these to be wholly discredited, regardless of whether their justification was transcendent, as in Platonism and Christianity, or immanent and located in the natural world. The former is, to be sure, no longer an option. But why should we also reject the latter, as Nietzsche would have us do? For it was precisely Aristotle's achievement to demonstrate that purpose and unity *are* inherent in Being, belonging to the natural world and finding a reflection in the social world of the *polis*. It was modernity's undue rejection of *this* heritage, Jonas suggests, an error from which neither Nietzsche nor Heidegger were exempt, that paved the way to absolute nihilism. Overcoming it requires, therefore, not a return to faith and revelation – which cannot hope to compete with secular developments – but rather the restoration, as far as is possible, of an Aristotelian philosophy of nature.

IV The age of technology

So far we have explored Jonas's analysis of modern gnosticism through the latter's relation to nihilism, which pertains, as we have seen, to the first two of gnosticism's three key principles. But this is only a part of its significance, as Jonas also believed that modern gnosticism was the root cause of both the ecological crisis and the development of potentially revolutionary biotechnologies. To understand why, we must firstly delve into the connection between modernity's metaphysical revolution and the development of modern technology. Only then can we turn to the third and final key feature of gnosticism – its eschatological dimension – which, in its modern manifestation, is intimately connected to technology's forward march.

Jonas's philosophical anthropology – which we shall look at in detail in Chapter 3 – affords the creation and use of technological artefacts a central place. Indeed, contrary to those critics who accuse him of technophobia (e.g. Ihde 1979: 132), Jonas straightforwardly claims that technology is 'integral to the human condition' (*IR*: 203). What concerns him is rather 'the triumph of *homo faber* [...] in the internal constitution of *homo sapiens*, of whom he used to be a subsidiary part' (*TPT*: 38). This shift can be brought out with reference to Aristotle, who as ever serves as a prime example of the Classical view. To flourish, Aristotle thought, we must identify and attain what is most valuable in life (1984d: 1140a–4a). Technology and invention, as means to ends, are therefore afforded a relatively low status among human affairs: they may help us achieve components of the good life, but they do not themselves constitute such a component. This could not be further removed from the spirit of modernity, which has elevated technological development 'to the position of [its] dominant and interminable goal' (*TPT*: 38). How, then, has this radical inversion occurred? To answer this question we must look, once again, to the scientific revolution.

According to the standard interpretation of Western intellectual history, '[m]odern science in its beginnings is the self-conscious renewal of the purely theoretical intention [...] of the ancient Greek will' (Melle 1998: 332). That is to say, the scientific enquiry emerging from the early modern era is – at least to begin with – thought to be identical to its Greek forebear in generating disinterested, theoretical knowledge of the world. Jonas, however, argues for quite the opposite interpretation. Hidden within the new conception of nature advanced by the work of Copernicus and Vesalius was, he says, a practical orientation unknown to Greek science – one that meant 'the technological turn later given to the speculative revolution was somehow in the cards from the beginning' (*PE*: 48).

The key difference between ancient and modern science is that from antiquity through to the Middle Ages speculative and practical knowledge were clearly distinguishable. The former kind of enquiry treated 'things unchangeable and eternal [...] which, being unchangeable, *can* be contemplated only', as in the disciplines of physics, astronomy and biology (*PL*: 189). The latter kind of enquiry dealt with 'the planned changing of the changeable', such as ethics, politics and the arts and crafts (*PL*: 189). While there was of course interaction between the two – biology informing medicine, for instance, leading to better health, which in turn contributed to the good life – theoretical enquiry remained logically distinct from practical affairs. The decisive change occurred with the modern conceptualization of nature as matter, devoid of teleology and explicable according to purely efficient causation. This vision contained 'manipulability at its theoretical core and, in the form of experiment, involved actual manipulation in the investigative process': the first of these characteristics leading to the second (*PE*: 48). The former concerns the mathematical interpretation of nature peculiar to modern science. Mathematics had certainly played an essential role in Pythagorean and Platonic ontology – indeed, legend has it that an inscription above the doors to Plato's Academy declared that only those who had mastered geometry were permitted entry. But according to Jonas, however, the geometric version of mathematized nature is crucially different to the *algebraic* model of modernity: the former was best equipped to map bodies in space, the movement of which remained explicable according to teleology, while the latter could also mathematically account for movement, and thus allow for its reduction to efficient causation (*PL*: 67, 92–5).

Once the movement of non-living objects had become fully calculable, it facilitated the second characteristic of modern science: experimentation. For only when movement could be mathematically accounted for could the measurable *manipulation* of the movement of those selfsame objects be carried out. Thus experimentation – for which the Greeks had little regard – was able to take its preeminent place in modern science. The significance of this development is itself twofold. Firstly, it entailed a partial dissolution of the ancient distinction between contemplation of nature (*theoria*) and the realm of practical intervention (*praxis*). Secondly, and crucially, by actively doing things to nature in the pursuit of empirical knowledge, instruments – technological means – could then take centre stage. From telescope to microscope to Large Hadron Collider, technology increasingly became the means by which scientific advances were made. It is for these reasons combined that Jonas boldly states that, from its inception, modern science was 'technological by nature' (*PL*: 198).

The consequence of these developments was that the previously linear and unidirectional relation between science and technology was transformed. No longer did scientific theory simply act as the bedrock upon which technological innovation occurred, but technological innovation also started to promote scientific discovery. Thus a circularity emerged in their relation: new scientific discoveries allowed for technological advances, which in turn propelled scientific research, which in turn generated novel technologies and so on, ad infinitum. Jonas calls this forward thrust the 'formal automatics' of modern technology (*TPT*: 36).

External influences then accelerated scientific technology's internal dynamic. Of greatest significance was the industrial revolution of the eighteenth and nineteenth centuries. New means of mass production saw technology put to the use of capitalism on a grand scale; as a result economic motives were introduced to the scientific enterprise, meaning that the tasks set for the sciences came to be 'increasingly defined by extraneous interests' (*FSI*: 16). In capitalist economies, where the industrial revolution began, these extraneous interests principally belonged to wealthy industrialists and the general consumer. Of critical importance is that by subjecting technological development to the operations of the market, everyone, and consequently no one, was made responsible for its direction. This responsibility became ever more diffuse as, over the course of the following two centuries, the capitalist mode of production spread worldwide, intensified further by the globalization of recent decades. Although it has tended to accommodate particular national traditions and circumstances, most countries – even those such as China, Cuba and Vietnam, which remain nominally communist – now collectively form a global technological-industrial civilization.

Crucially for Jonas, this increased scale of technological development brought about an increase in its autonomy, fuelled by the dynamic of supply and demand. According to Jonas this is partly a matter of advertising: 'industry does not so much strive to *fulfill* human needs as to *generate* human needs', illustrated by the fact that 'nobody ever dreamed of most of the things that progress keeps offering for our consumption' (*CR*: 216; emphasis added). But there is, in addition, a more pervasive and objective aspect to the pull of development. Technologies themselves 'suggest, create, even impose new ends [...] simply by offering their feasibility' (*TPT*: 36). We are, in other words, encouraged to adopt novel technology because prior technology makes it desirable, on both an individual and societal level. Jonas continues: 'Technology thus adds to the very objectives of human desires, including objectives *for technology itself*. The last point indicates the dialectics or circularity of the case: once incorporated into the socioeconomic demand diet, ends first [...] generated by technological invention become necessities of life and set technology the task of further perfecting the means of realizing them' (*TPT*: 36; emphasis added). The automobile is a good example. Cars and trucks were not isolated technological inventions, but inspired tarmacked roads, motorways, petrol stations, speed cameras, traffic lights, new laws, taxes and insurance, greater opportunities for trade and so on. All became desirable when accommodating the new technology into our socio-economic life, and any of *these* technologies or practices may serve to facilitate others.

This dialectic lies at the heart of Jonas's theory of technology: that the positive feedback loop of scientific-technological innovation and industrial capitalism strengthens in line with our reliance upon ever more technology.

The development of modern technology has, then, long since become a *systemic* phenomenon. Of course, this system is neither deterministic nor totally inescapable: each of us has the ability to renounce certain technologies and practices, from the luxurious like air travel to the prosaic such as owning a car. But it is only really feasible for individuals to opt out of the less essential aspects of modern life: day to day, we still have to live somewhere, work, travel, eat, drink, wash, clothe ourselves and so on, all within the societal parameters in which we find ourselves. 'Dropping out' may be an option for some of us, but surely not all. In short, '[A] style of life has been established which one somehow has to join […] or else one simply cannot exist in the society as it has come to be' (*CR*: 216). Thus we each play our part in the socio-economic whole and contribute, however indirectly, to the forces driving technological development.

By characterizing contemporary technology as out of control, and tracing its origins to science and mathematics, there are clear parallels between Jonas's philosophy of technology and that of Heidegger. Curiously, however, Jonas very rarely mentions this part of Heidegger's thought and neglects to acknowledge his debt to it. The most notable exception occurs in his *Memoirs*, where Jonas states: 'We need a new ethics for the age of technology, one that confronts the challenges of our era. Heidegger, for one, recognized this need, and attempted to take it on, though what he has to say on the subject […] seems to miss the point completely' (*M*: 203). The differences between Jonas's philosophy of technology and Heidegger's are in fact fairly subtle, yet also essential for understanding his philosophical response to modern gnosticism. We shall, therefore, briefly outline the three key differences between their analyses.

The first pertains to the origin of technology in mathematics. For Heidegger, as well as Jonas, the particular character of mathematics was key to understanding modern technology. Heidegger held, however, that manipulability is not only particular to the role mathematics plays in modern science but in fact its essence *as such*, mathematics consisting only of 'a kind of grasping and appropriating' of beings (1977: 251). Because Jonas distinguishes between sorts of mathematics and Heidegger did not, their theories accordingly differ regarding the pivotal historical moment in the genesis of modern technology. As we have seen, for Jonas this moment is the scientific revolution of modernity, which rested on an algebraic mathematization of nature. For Heidegger, however, the emergence of modern technology through science is nothing less than the *culmination of*

Western metaphysics. Throughout our long history, being has been construed as some-*thing* that grounds all other beings: at various points this role has been played by the God of Abrahamic religion, the Forms of Plato and, eventually, the laws of nature mathematically described by the natural sciences. The latter, giving rise to modern technology, is therefore only the last and purest expression of the 'productionist' understanding of being inherent to Western thought. It is – to say the least – an audacious argument, and one that Jonas sensibly avoids making.

The second key difference concerns the extent to which modern technology has escaped human control. Jonas's theory of technological autonomy is rooted, as we have seen, in a concrete analysis of technology's interactions with other activities, namely scientific practice and globalized industrial capitalism. Thus technological development, while beyond the control of any one individual or nation state, is still in principle subject to collective human willpower.[6] Heidegger, by contrast, held that 'technological advance will move faster and faster and cannot be stopped', as '[t]hese forces […] have moved long since beyond [man's] will and have outgrown his capacity for decision' (1966: 51). Having located the origin of technology in our productionist metaphysics, Heidegger regarded any attempt to control or assert our will over technological development as simply another manifestation of the very same drive to grasp and appropriate beings.

The third and final difference between the two philosophies pertains to how we might respond to out of control technology. For Jonas, the way to restore technology to its rightful place – as a means, rather than an end – is with 'moral reason' (*IHJ*: 368). Through public policy, underpinned by ethical analysis, Jonas hopes that collective action might be able to set appropriate boundaries on technological development, though he admits that this is 'a very weak and frail hope' (*IHJ*: 368). Heidegger, unsurprisingly, had no such hope at all. In his interview with *Der Spiegel* Heidegger famously claimed that 'philosophy will not be able to bring about a direct change of the present state of the world. This is true not only of philosophy but of all merely human meditations and endeavors. Only a god can still save us' (1990: 56–7). All we can do is seek 'releasement' from technological appropriation through poetry and post-metaphysical thinking (2010b: 79). Then one must simply await the arrival – or fully comprehend the absence – of the gods.

Heidegger's canonical reflections on technology undoubtedly constitute part of his huge influence on Jonas, albeit one that went largely unacknowledged by the latter. However, Jonas's fidelity to ethics and reason ultimately sets him apart from Heidegger, in this area as in so many others. Identifying the source of unbridled technology as a particular kind of mathematized nature, intertwining with science and industrial capitalism, Jonas avoids the trap of a blanket criticism of science and mathematics into which Heidegger fell. Jonas is a staunch critic of scient*ism*, no doubt, but his work reveals genuine affection and respect for the sciences proper. Moreover, as his diagnosis remains within the realm of human thought and action, so too does his envisioned solution of a reasoned moral response. In what can only be a rebuke of Heidegger, Jonas tells us: '[T]he human mind alone, the great creator of the danger, can be the potential rescuer from it. No rescuer god will relieve it of this duty' (*MM*: 54).

V The Baconian ideal

We have now explored all aspects of Jonas's critique of modern gnosticism except one: gnosticism's eschatological dimension. Eschatology is traditionally the part of theology concerned with life after death, both with regard to the fate of the individual soul and that of the world as such. According to Christian doctrine, for instance, after the death of the body the soul is bound for either heaven (sometimes via purgatory) or hell. And eventually, following the Second Coming of Christ and Last Judgement, the universe itself stands to be renewed, with heaven realized on earth. The Gnostics, to give another example, believed in a rather more solipsistic eschaton. As discussed, according to the religion's tenets, the individual soul may in death escape the malign physical world and return to the world of light, provided that in life it had attained *gnōsis*. In the vast majority of Gnostic texts, this constitutes the religion's entire eschatological doctrine: that is to say, the fate of the world as such is overlooked in favour of the fate of the individual soul (*GR*: 45). For the Gnostics neither collective salvation nor judgement awaits us at the end of history.

In spite of the differences between their ideas of individual and collective fate, the Christian and Gnostic eschatologies share a fundamental structure. Firstly, note that in both religions salvation is to be found in the soul's escape from the natural world. In Gnostic eschatology this happens on an individual basis immediately after death; in the Christian version the route is more circuitous, ending on a renewed world free of suffering and death. Nevertheless, in both cases salvation entails an *overcoming of nature* – which, as we have seen, is devalued in their gnostic ontologies – to arrive at a state of paradise. The second commonality is that both eschatologies privilege those who in life observed Holy Writ, while those who worshipped false idols are excluded from utopia. Thus by the manner in which we live each of us makes a choice between possible futures.

How, then, does this last aspect of gnosticism manifest in modernity? As before, what was originally a transcendent principle of existence is in modernity made immanent, this time in the form of techno-utopianism. The techno-utopian agenda follows almost as a matter of course from the previous components of modern gnosticism, hence its concluding place in our account. Having divorced mind from nature, devalued the latter and made unprecedented technological development possible, modern gnosticism's complementary eschatological claim is that mind must *transform* nature – including human nature – to create utopia here on earth. In this is a perfect mirror image of earlier gnostic beliefs. As Eric Voegelin says, drawing on Jonas's work, 'progressivism, positivism, and scientism' are essentially materialist translations of the symbolic structure of Christian eschatology, constituting an 'immanentization of the eschaton' (1952: 163–4). Whereas the Christians and Gnostics sought salvation in leaving this world for the next, therefore, techno-utopianism is instead oriented towards the technological transformation of the given world into a 'state of perfection' (1952: 120).

This techno-utopianism was precisely Descartes's hope for the modern sciences. In his *Discourse on the Method of Rightly Conducting One's Reason and of Seeking Truth in the Sciences*, Descartes suggested that 'instead of the speculative philosophy taught in the Schools' – meaning that of Aristotle and Aquinas – 'a practical philosophy can be found', which would make us 'masters and possessors of nature' (1968: 78). With his mechanistic understanding of life and dualistic metaphysics, Descartes made significant and sophisticated contributions towards achieving that end, as we have seen. However, the core of the new practical philosophy had already been developed across the English Channel by Jonas's other bête noire: the statesman, scientist and philosopher Sir Francis Bacon.

Bacon's ideal consisted of a method and a goal that together encapsulate the modern eschaton. For his method Bacon radically departed from the established Aristotelian-Scholastic form of scientific enquiry. In the texts collectively known as the *Organon*, Aristotle had stressed the importance of syllogistic reasoning, prioritizing this over induction and empirical observation. Two millennia later in his *Novum Organum* – literally, the *New Organon* – Bacon argued for the very opposite, subordinating syllogistic reasoning to what we today call the scientific method: observation, experimentation and inferential reasoning. In doing so Bacon hoped to topple the 'idols' that he thought clouded human judgement and led us astray in the search for truth ((1620) 2000: 40). Among what he called the 'idols of the tribe' were the prejudices of the human mind and the limitations of our senses: the former habitually imposing order and purpose on nature where there is none, the latter preventing us from directly observing nature's most fundamental elements and processes ((1620) 2000: 41). The 'idols of the marketplace', by contrast, were everyday discourse and mainstream opinion. These he took to be 'the biggest nuisance of all' due to their ubiquity and the fact that they were tailored to the 'capacity of the common man', rather than that of the man of learning ((1620) 2000: 48).

The end of the Baconian ideal was as equally radical as its means. From Aristotle to Aquinas theoretical contemplation was thought to play a key role in attaining the good life, and both figures cherished scientific enquiry for that very reason. Bacon, by contrast, saw the value of his new method as its potential to serve more instrumental needs: science must strike out in this new direction, he says, 'so that the mind may exercise its right over nature' ((1620) 2000: 6). The imperative of domination is perhaps most famously formulated in his claim that 'those two goals of man, *knowledge* and *power*, [...] are really come to the same thing' ((1620) 2000: 24). If humanity is able to control nature through scientific knowledge – so the argument goes – nature might be used against itself by alleviating the limitations, dependencies and weaknesses inherent to humanity's biological constitution.

The radicalism of Bacon's vision is made clear in his *Valerius Terminus*. In the Judeo-Christian creation myth humanity had been depicted as having dominion over the Garden: according to Jonas, Adam's ability to name and thus symbolically order every living creature therein 'foreshadowed [humanity's] coming mastery over nature' (*PL*: 173). Invoking the story of Genesis, and the connection between knowledge and power that it depicts, Bacon claims that the 'true ends of knowledge' are the 'command' of nature and the 'restitution and reinvesting [...] of man to the sovereignty and power (for whensoever he shall be able to call the creatures by their true names he shall again command them) which he had in his first state of creation' (1984: 42). For Bacon such command ranged all the way from 'the meanest mechanical practice' to 'immortality (if it were possible)' (1984: 42).

The connections between Baconianism and gnostic eschatology should by now be clear. In the former the scientific method takes the place of *gnōsis*, while those who possess it are the men of reason rather than those beholden to the idols of the tribe and marketplace – the latter reminiscent of the somatics. And the decisive parallel, of course, is that Bacon quite literally held utopia to be the historical *telos* of modern scientific technology. Here we have an immanentized version of the overcoming of nature that gnosticism in general promotes as the means of salvation; in Bacon's version, the overcoming simply consists of a physical transformation. Even more so than Descartes, then, Bacon articulated the modern eschaton: an ideology of progress that is neither a 'gloss' on nor 'a mere option' of technological development, but rather technology's '*modus operandi* as it interacts with society' (*TPT*: 35).[7] Hence we moderns are, as Jonas says, 'Baconians rather than Aristotelians with respect to the kingdom of man' (*OF* V: 63).

Now, we have already seen how the first two aspects of modern gnosticism led to the nihilism that presently afflicts us. Having since looked at the third and final facet, the connection between modern gnosticism and the twin concerns of environmental ruin and biotechnological developments may be explained. Recall that modernity's metaphysical revolution symbolically began in 1543 with the publication of key works by Copernicus and Vesalius. Jonas observed that together these texts represented the emerging materialist world view: the former at the macroscopic level of the cosmos, the latter at the microscopic level of the organism. Once construed in mechanistic terms their technological manipulation was made possible; with the accompanying eschatological promise of utopia, and no welfare or interests but our own to take into account, it then also became *desirable*. Practically this entailed, as Bacon famously put it, 'relief of man's estate' through the extraction of natural resources and the medical conquest of illness, weakness and even death itself ((1605/1627) 1906: 39).

It goes without saying that in many respects the Baconian ideal has proven a resounding success. In the industrialized world, at least, average life expectancy has dramatically risen, terrible diseases have been widely cured and the necessities of life made readily available to most citizens. Overall, the average Westerner's life is unimaginably easier than in Bacon's day – even if this is partly because the West extensively relies upon the cheap labour of the Global South. And yet the very same process of development has also led to climate change, mass species extinction, the desertification of landscapes, pollution, overpopulation, ocean acidification and, finally, biotechnology, the prospective use of which poses radically new ethical challenges. Clearly, Bacon himself could not have foreseen the harms and dilemmas that would arise in the course of realizing his vision, but we have for decades now been aware of an 'apocalyptic perspective calculably built into the structure of the present course of humanity', which Jonas calls the 'ominous side of the Baconian ideal' (*IR*: 140–1).

The environmental issues mentioned – which may without exaggeration be collectively referred to as an ecological crisis – typically pertain to 'slow, long-term, [and] cumulative' threats, and are therefore easily overlooked (*IR*:

ix). Worse, we have a vested interest in turning a blind eye, as all arise from the higher standards of living to which modern humanity actively strives. The scientific data, however, are clear as to the changes occurring and the problems posed on both the global and the intergenerational scales.

This is perhaps best illustrated by climate change. The most recent Assessment Report from the UN Intergovernmental Panel on Climate Change (IPCC) found that '[w]arming of the climate system is unequivocal, and since the 1950s, many of the observed changes are unprecedented over decades to millennia. The atmosphere and ocean have warmed, the amounts of snow and ice have diminished, sea level has risen, and the concentrations of greenhouse gases have increased' (2013: 4). The negative consequences will then reach into the distant future. The IPCC claim that a 'large fraction of anthropogenic climate change resulting from CO_2 emissions is irreversible on a multi-century to millennial time scale' (2013: 28). Even if carbon dioxide emissions were immediately and dramatically reduced, thus limiting the global mean temperature increase to 2°C above preindustrial levels, '[m]ost aspects of climate change will persist for many centuries' (2013: 27) accompanied by the 'considerable' risk of extreme weather events and harms to unique and threatened natural systems (2014: 14). If our present rate of emissions continues unabated, however, leading to a global mean temperature increase of 4°C or more above preindustrial levels, the likely effects include 'severe and widespread impacts on unique and threatened systems, substantial species extinction, [and] large risks to global and regional food security' (2013: 14). In the worst-case scenario, mutually reinforcing climactic tipping points could generate a positive feedback loop of extreme warming, radically diminishing the planet's habitability (Steffen et al. 2018: 8256). Whether we will act in time to prevent this is an open question.

Such is the state of affairs at the macroscopic level of the climate. At the microscopic level, the novel biotechnological power to precisely manipulate organismic development and the evolution of species raises a host of unprecedented ethical concerns. From the first cloning of non-human life in the 1950s, followed by genetic engineering in the 1970s, through to the synthetic biology and prospect of 'designer babies' today, the genome has become the direct object of our control.[8] The accuracy with which genes can now be targeted for modification, particularly since the development of the CRISPR/Cas9 tool, dwarfs that of traditional methods such as cross-pollination and animal husbandry. This growth of power does not, to be clear, necessarily reflect a transition from ethical to unethical methods, but it does drastically increase the urgency of establishing the acceptability of their purpose. On what – or whom

– should we use such technologies? To what ends? Should their use be restricted to curing disease – the noble and time-honoured task of medicine – or may we also explore their creative capacity and eugenic potential? If we did pursue the latter course of action, what would this say about us?

Unfortunately, the technologies in question are developing faster than such questions can be answered and global standards set. For a while a fragile international consensus held against bringing genetically engineered human embryos to term, at least until any potential risks were better understood. Then, in November 2018, this settlement was dramatically overturned by the maverick Chinese scientist He Jiankui. On the eve of the Second International Summit on Human Genome Editing, it was revealed that He had indeed brought to term two genetically engineered children, modified to carry a possible resistance to HIV. Although the revelation was met with near-universal condemnation, the floodgates may now be open and prove impossible to close. To stem the tide we would have to find the will to resist 'the ruling pragmatism of our time', which 'will let no ancient fear and trembling interfere with the relentless expanding of the realm of sheer thinghood and unrestricted utility' (*PE*: 142).

And so modern gnosticism has led humanity into a situation without historical precedent. Late antiquity may have witnessed the rise of the Gnostic religion and the demise of Classical civilization, but the future of humanity and life on earth were never in doubt. Today, however, we have no such assurances. Ecological degradation could make swathes of the planet hostile to life, while the indiscriminate use of biotechnology threatens to violate our unique human dignity. Worse, any attempt to formulate a collective response is paralysed by a lack of binding norms: humanity's moment of giddiest technological might coinciding with its moment of deepest nihilism.

We are left, therefore, with no choice but to confront and overcome modern gnosticism. But what would its adversary look like? Recalling gnosticism's most positive trait – the enrichment of the person – Jonas suggests that although the 'dualistic rift' must be avoided, enough of its 'insight must be saved to uphold the humanity of man' (*PL*: 234). His magnificent attempt to strike this balance is based on the following anti-gnostic principles – one ontological, one axiological and one ethical.

1. Humanity is the pinnacle of life's tendency towards freedom.
2. All living beings are ends-in-themselves, but human beings alone have personal dignity.
3. Humanity has a responsibility to ensure its own continued existence on a habitable earth.

In this way Jonas hopes to reunite ontology and ethics, the marriage of which was 'the original tenet of philosophy' (*PL*: 283). Staying true to the Greek legacy, he tells us that 'Being, in the testimony it gives of itself, informs us not only about what it is but also what *we owe to it*' (*MM*: 101; emphasis added). The remainder of this work will seek to demonstrate, as far as is possible, the validity of this declaration, starting with the first and most fundamental component of Jonas's philosophical system: humanity's true place in nature.

2

The philosophy of life I
The organism

I Dualism, materialism, integral monism

Jonas's analysis of gnosticism led us to identify modern materialism as the root cause of our contemporary plight. Denying purpose and order to the cosmos, materialism denies any value to nature except that of meeting humanity's needs – the relentless expansion of which has, through the medium of modern technology, led to the ecological crisis and our self-objectification through biotechnology.

This is reason enough, we would suggest, to search for a superior ontology: one that can do justice to the purposes of non-human nature while making room for humanity's uniqueness. Jonas's theory of Being sets itself precisely this lofty goal, and its reconstruction is the purpose of both this chapter and the next. To this end we shall draw extensively on the two texts in Jonas's oeuvre that offer a detailed presentation of his ontology: *The Phenomenon of Life* and *Organism and Freedom*.[1] The former text is typically thought of as Jonas's definitive statement on issues of ontology and existentialism, and with good reason. But it is not *quite* comprehensive, as two issues, in particular, receive insufficient treatment despite their importance for his philosophical system: the interpretation of metabolism, and the analysis of animal life. It is the second text, *Organism and Freedom*, that fully addresses these topics, and so we shall place greater emphasis on it than Jonas's commentators have to date.

In order to understand how Jonas arrives at his alternative ontology, we must begin with his account of the limitations of materialism. We described in Chapter 1 how modern materialism originated in the vision of nature presupposed by the new physical sciences, which led to its philosophical codification as *res extensa* in Descartes's substance dualism. Of course, Descartes gamely attempted to reconcile the *res extensa* with the immediate and indubitable evidence of consciousness, leading to him to posit to the existence of a second, distinct substance: the *res cogitans*. This 'bifurcation of nature', as Alfred North Whitehead had it, preserved the domain of mind from mechanistic reductionism, but only at the considerable cost of denying subjectivity to non-human life and failing to satisfactorily account for the intimate connection between the mind and the body (1920: 32). It is precisely this gnostic tendency that Jonas wishes to overcome with a new, 'post-dualistic' ontology (*PL*: 16).

At first glance either idealism or pure materialism, the competing monistic ontologies that arose in Cartesianism's wake, would fulfil this criterion. Yet despite appearances this is not in fact the case. Each ontology represented just one half of the Cartesian bifurcation of nature – idealism the *res cogitans*, materialism the *res extensa* – and attempted from its vantage point to account for the domain of Being belonging to the other. Precisely because of this inherent limitation both ontologies were destined to fail, as we can briefly but revealingly demonstrate with reference to one phenomenon in particular.

Idealism was undoubtedly the more sophisticated of the two ontologies, and is associated with the great figures of Berkeley, Kant, Hegel and Schopenhauer. Philosophically retreating into the subject, Berkeleyan idealism took the world of experience to be simply an idea held in the mind of its perceiver, the latter itself sustained in continual existence by the mind of God. Following Kant's critical turn, however, idealism sought to explain the world as a phenomenal realm constituted by the transcendental subject. In either form, according to Jonas, idealism's attribution of ontological primacy to the subject necessarily led to the reappearance of the most intractable problem of Cartesian substance dualism: namely, how we are to make sense of the subject's embodiment. For if the body 'is but one among the "external ideals" ("cogitations") of consciousness', as idealism must suppose, then the fact of 'its being the body *of* this consciousness, its being *my* body, my extended I and my share in the world of extension, is not understood' (*PL*: 18).

In perfect parallel, materialism ceded the terrain of mind and so found itself having to account for the immediate evidence of consciousness on impoverished terms. At best this entailed a theory of mental emergence, according to which the causally efficacious mind is located in particular configurations of bare matter. As Jonas points out, however, this is but a 'valiant attempt' to hold on to both pure matter and the power of consciousness, while refusing to admit that *something* of the causal power of mind would have to already reside in matter in order for particular configurations of the latter to give rise to the former (*IR*: 67). At its worst, materialism sought to account for the mind as a mere epiphenomenon of brain states, lacking any power over the mechanistic body. This theory Jonas rightly regards as a 'logical absurdity', since mechanistic processes are 'entirely foreign to "meaning" and "truth"' which are both presupposed in making the case for epiphenomenalism (*PL*: 129). The very act of arguing for the reduction of mind to a mere shadow of neural activity is, therefore, a performative contradiction.

Our account of the structural limitations of idealism and materialism, though brief, points towards the way in which both are to be overcome. Neither ontology could account for the domain of Being belonging to the other, and in both cases this failure circulated around the lived body: idealism could not explain the essential corporeality of the subject, while materialism could not coherently account for the immediate evidence of bodily consciousness. In this aporia lies the clue as to how to proceed. For what it reveals is that *the living body straddles the purported division of mind and matter*, proving fundamentally irreducible to one side or the other. Hence only an 'integral' monism, committed to the unity of matter and mind, could possibly resolve 'the problem of life, centred in the problem of the body' (*PL*: 19). This Jonas sets as his task: an explication of the living body on the terms of a non-reductive monism, hoping with such a theory of Being to banish the spectre of gnostic dualism that has haunted modern Western thought up to the present day.

How, then, is such an ontology to be constructed? Jonas holds that an integral monism would take the living body as not only its cardinal theme but also its methodological point of departure. The reason for doing so is that '*my* body [...] is, in its immediacy of inwardness and outwardness in one, the *only* fully given concrete of experience in general' (*PL*: 24). Jonas constructs his ontology, therefore, via a phenomenological account of embodiment that draws on scientific findings to identify our place in nature – or, as he puts it, an '"existential" interpretation of biological facts' (*PL*: xxiii).

In his method, if not the detail of his ontology – which owes more to Aristotle – Jonas once again demonstrates a profound debt to Heidegger.[2] Although Jonas departs from it in important ways, as we shall see, his approach is heavily influenced by Heidegger's early existential phenomenology, and in particular the monumental *Being and Time*. Heidegger himself, however, owed a great deal to the work of his own teacher: the founder of phenomenology, Edmund Husserl. To understand Jonas's particular version of the phenomenological method we shall, therefore, start with Jonas's rejection of Husserl before turning – and not for the last time – back to Heidegger's existentialism.

II The phenomenological approach to organismic being

In 1921 Jonas was drawn to Freiburg by Husserl's reputation as the greatest thinker of his day. Jonas always viewed him with a deep respect, and following Husserl's death in 1938 praised him highly in an obituary broadcast on public radio from Jerusalem. Philosophically speaking, however, Jonas regards Husserl's thought as belonging squarely to the idealist tradition discussed.

At the turn of the twentieth century Husserl had founded the philosophical movement known as phenomenology, which seeks a fundamental account of the way in which phenomena appear. Beyond this very broad aim, however, there is little consensus as to what the method necessarily consists of. For Husserl the rigorous study of phenomena could only be carried out by suspending belief in their independent existence, allowing one to focus instead on how the world appears 'precisely as it is for me' (1999: 21). This meant, as Jonas notes, that philosophy for Husserl amounted to the 'science of consciousness', with study of the objective world left to the natural sciences (*EH*: 15). But philosophy, if it is to transcend Cartesian dualism, has to go further and establish 'in what sense and to what extent we are enmeshed in the processes of nature' (*IHJ*: 345). Husserl's neglect of this vital task makes his version of the phenomenological method incapable, according to Jonas, of overcoming modern gnosticism.

It has to be said that Jonas's characterization of Husserl's thought is unfair, however, as in his final years Husserl both engaged with the natural sciences and stressed the 'completely unique ontic meaning of the body' (1970: 106–7). But it is true to say that the focus of Husserl's middle period, which coincided with Jonas's time studying under him, was devoted to the structure of consciousness and its constitution of the world, thereby representing a continuation of transcendental idealism. By contrast, the existential version of phenomenology developed by Heidegger, who was at that time Husserl's assistant, examined the human condition from its situation amid the world of people and things. In Chapter 1 Jonas's relation to Heidegger's existentialism was explored in terms of the gnostic aspects he rejects: namely, its implicit adherence to the dualism of humanity and nature, and its resultant nihilism. Despite this, Jonas takes inspiration from certain key themes of Heidegger's existential analytic to develop his own philosophy of organismic being, *against* Heidegger's gnosticism. To establish how he does so, let us delve deeper into Heidegger's thought.

As we have seen, the majority of *Being and Time* was devoted to a phenomenological analysis of the being of *Dasein*. Crucially, Heidegger's investigations did not look 'inwards' to the structure of the transcendental subject, as Husserl's had, but out to the world. This is for the simple reason that *Dasein* is not first and foremost a consciousness, distinct from the physical world of objects it encounters, but rather fundamentally *engaged* with beings that appear in its environment and overarching world. In Chapter 1 we saw that Heidegger described the being of *Dasein* as absorption in projects both informed by the culture into which it is thrown and oriented towards the future. We shall now sketch out precisely what Heidegger took *Dasein*'s projects to consist of.

Heidegger observed that *Dasein* is primarily and for the most part engaged with beings in its environment instrumentally. Heidegger called these engagements *Dasein*'s 'dealings', a mode of absorption that encompasses everything from opening a door to counting loose change to tilling soil (2010a: 69). These beings are all encountered as ready-to-hand means to various ends, and the more successfully they allow for the realization of those ends, the more the beings in question recede into the background of our attention. Such a being only comes to be an object of study when it breaks, malfunctions or otherwise throws us off, and suddenly commands our attention in the form of a problem. The latter type of absorption, of reflection and questioning, constitutes what Heidegger called 'cognition' of things (2010a: 71). It is the kind of engagement that underpins scientific and theoretical enquiry, and is, he claimed, merely derivative of and secondary to our absorbed dealings.

Of course, we do not always engage with beings in just these two ways. Most obviously, we are absorbed in the lives of other people, who appear not as things but as *Dasein* like ourselves: a mode of absorption Heidegger called 'concern [*Fürsorge*]' (2010a: 118). Heidegger claimed that concern is not a matter of whether we are, in fact, in the company of other people at a given time, as even in the total absence of others we are conscious of our isolation and perceive it as a privation. Indeed, Heidegger even claimed that of its various possible 'ways of concern' – namely, '[b]eing for-, against-, and without-one-another, passing-one-another-by, [and] not-mattering-to-one-another' – *Dasein* 'initially, and for the most part, lives in the deficient modes' (2010a: 118). This judgement may well be pessimistic, but the broader point stands: whether we are physically with other people or not, 'for' them in our projects or not, *Dasein* is just as frequently engaged with others at is it with things.

We now reach what is perhaps the crucial moment in Heidegger's existential analytic of *Dasein*, which is openness to the world that we exist in. Heidegger firstly distinguished between the world (*Welt*) and the environment (*Umwelt*). The latter represents our surroundings, which are not encountered in the form of distinct objects subsequently synthesized by consciousness, but are rather given as a whole towards which one is oriented in absorption. 'World', by contrast, has several meanings too nuanced to individually recount here, but suffice it to say that for Heidegger the world *ontologically* understood is neither the earth nor the totality of objects constituting the physical universe, as we might assume. Rather, the world is the referential whole of significance that we encounter in our environments. An example is a workshop, where the activities absorbed in – hammering a nail, for example – point towards an entire nexus of needs belonging to oneself and others: building a cabinet to store clothing, within a dwelling where *Dasein* takes shelter, cohabiting with one's family to form a private sphere within the public realm. As with the environment, the world is therefore encountered through our dealings, and for this very reason Heidegger argued that *Dasein* can only be understood as 'being-in-the-world' – the hyphenated formulation emphasizing the fundamental unity between *Dasein* and the world in which it is absorbed (2010a: 57).

All of the above, in conjunction with existence and the account of thrownness, fallenness, and being-toward-death given in Chapter 1, constitute the being of *Dasein*: a '*being-ahead-of-itself-in-already-being-in-a-world*' (2010a: 185). This structure Heidegger called 'care', and he took it to be the 'existential *a priori*' of any kind of subjective attitude or position that it is possible for *Dasein* to take; thus *Dasein*, '*ontologically* understood, is care' (2010a: 187, 57).

With the being of *Dasein* identified as care, which was itself shown to be an inextricable 'entanglement' in the world, Heidegger's existential analytic of *Dasein* circumvented the standard philosophical account of a rational subject separable from a realm of objects (2010a: 221). In exactly this radical achievement lies its greatness. And viewed in light of Heidegger's statement that *Being and Time* would pave the way for a 'phenomenological destruction of the history of ontology', with the Cartesian *res cogitans* being a prime candidate for such treatment, it seemed that Descartes's dualistic legacy had finally been overcome (2010a: 37).

The extent to which this was not the case, however, is made clear when we examine Heidegger's treatment of animals. Any truly post-dualistic ontology would have to account for the connection between humanity and the world of nature to which we belong, yet if we take animals as our representative of the latter then it quickly becomes apparent that Heidegger failed in this regard. A survey of his essays and lecture courses written from the beginning of the 1920s to the middle of the 1930s shows that Heidegger struggled with the question of *Dasein*'s connection to non-human life. Since *Dasein* can only be understood as being-in-the-world, Heidegger's investigation focused on whether animals engage with beings in a way that indicates openness to the world (*Welt*), or, failing that, openness to their immediate environment (*Umwelt*). If animals could be shown to have access to the world then they would necessarily count as *Dasein*; access to the environment alone would at least guarantee partial status as such.

At his most generous, in a 1925 essay, Heidegger suggested that the life of the animal 'is that kind of reality which is in a world and indeed in such a way that it has a world. [...] [W]e miss the essential thing here if we don't see that the animal has a world' (2002: 163). Here his affirmation of the worldhood of animals was unequivocal. By 1929, however, he had revised his earlier assessment and instead conceived of animals as 'poor in world' (1995: 176). In his most sustained and penetrating analysis of the issue, Heidegger drew on biological studies to argue that animal behaviour indicates a captivation by beings, meaning that although animals are open to beings in their immediate environments, this does not extend into a world proper. Eventually, in 1935, following his turn away from existential phenomenology, Heidegger denied even this degree of openness to animals. The 'animal has no world, nor any environment', he claimed (2000: 47).

Why could Heidegger not arrive at a satisfactory account of non-human life? Why did his early sympathies for animal being-in-the-world collapse into an outright dismissal of the idea? Jonas argues that it was because Heidegger had from the beginning overlooked the corporeality of *Dasein* and its connection to nature. In *Being and Time* 'the body had been omitted and nature shunted aside as something merely present', meaning that Heidegger ignored the extent to which embodiment, encountered both phenomenologically and biologically, is in fact *constitutive* of *Dasein*'s being (*WPE*: 31). More specifically, Jonas observes that in Heidegger's early thought there 'was no mention of the primary reason for having to care, which is our corporeality, by which we – ourselves a part of nature, needy and vulnerable – are indissolubly connected to our natural environment' (*WPE*: 31). And yet more concretely, he wonders why care was never traced back 'to concern for nourishment, for instance – indeed, to *physical* needs at all' (*MM*: 47). The reason for this, as discussed in Chapter 1, was that Heidegger inherited the deficient understanding of nature belonging to modernity, which was impossible to reconcile with his rich account of *Dasein*. This ultimately gave rise to the gnostic dimension of Heidegger's thought, including an inability to conceive of an existential commonality between humanity and other forms of life. 'Somehow', Jonas suggests, 'German philosophy with its idealistic tradition was too high-minded to take this into account' (*MM*: 47) – an observation borne out by Heidegger's startling claim that 'the essence of divinity is closer to us than [...] our scarcely conceivable, abysmal bodily kinship with the beast' (1977: 206).

Although Heidegger broke with modern ontology in certain key ways, then, his thought was hampered by an all-too-modern neglect of the body that connects us to living nature.[3] Crucially, however, the failure to link *Dasein*'s being to its embodiment was not the result of an inherent limitation of existential phenomenology, but rather Heidegger's particular formulation of it. For this reason Jonas is able to take up key aspects of it for the purpose of finally overcoming the gnostic division between humanity and nature. This programmatic intent is set out on the very first page of *The Phenomenon of Life*, which I shall quote from at length:

> Contemporary existentialism, obsessed with man alone, is in the habit of claiming as his unique privilege and predicament much of what is rooted in organic existence as such: in so doing, it withholds from the organic world the insights to be learned from awareness of self. [...] Accordingly, the following investigations seek to break through the anthropocentric confines of idealist and existentialist philosophy as well as through the materialist confines of natural science. In the mystery of the living body both poles are in fact integrated. The great contradictions which man discovers in himself [...] have their rudimentary traces in even the most primitive forms of life, each precariously balanced between being and non-being, and each already endowed with an internal horizon of 'transcendence'. We shall pursue this underlying theme of all life in its development through the ascending order of organic powers and functions [...] – a progressive scale of freedom and peril, culminating in man, who may understand his uniqueness anew when he no longer sees himself in metaphysical isolation. (*PL*: xxiii)

The perceptive reader will note that Jonas seeks to trace 'much', not all, of Heidegger's existential analytic back to the living body, which is only to be expected given his rejection of the latter's gnosticism. Unfortunately Jonas does not straightforwardly state which aspects he rejects and which he incorporates into his philosophy of life, meaning that a degree of creative interpretation is required. Proceeding with this caveat in mind, we may say on the basis of Chapter 1 that Jonas rejects Heidegger's accounts of thrownness, fallenness and authenticity on the grounds that these are not universal to *Dasein*, but in fact culturally and historically contingent. And in Chapter 3 we shall see that Jonas agrees with Heidegger that existence – *Dasein*'s understanding of itself in its being – is indeed something that belongs to humans alone. In the present chapter, however, we shall follow Jonas in tracing the remainder of the care-structure – that is, being-ahead-of-itself-in-already-being-in-the-world – back to organismic being as such. The resulting ontology is comprised of two key ideas: on the one hand is a description of the individual 'living, feeling, striving organism', and on the other is an argument concerning the conditions of possibility for the evolutionary emergence of such beings (*PL*: 12).[4] As the former is more fundamental to his enquiry, we shall give priority to it in what follows.

Jonas's reconciliation of 'the organic facts of life' with 'the self-interpretation of life in man' begins with a phenomenology of embodiment (*PL*: 6). The first thing to note is that embodiment is here understood in a specific way. For Jonas embodiment does not mean 'having' a body, insofar as this turn of phrase implies that the self is still distinct from the body that it has. On the contrary: it means that the self *is* the body. This identification of body and self holds even when we scientifically objectify ourselves in thought, regarding our bodies as though from the outside. For in such acts of self-objectification my body is carried with me in implicitly remaining the position from which things appear as external to me at all. In other words, I can conceive of my body as an external object only because I have first of all encountered such objects through it. Embodiment means for Jonas, therefore, not that I have a body but that I am a body, and uniquely *this* body, 'to the tips of my fingers' (*PL*: 18).

Jonas argues that an embodied phenomenology has distinct explanatory advantages over Husserl and Heidegger's quasi-Cartesian versions, which we may illustrate with the following example. Consider the statement 'I am hungry'. Jonas claims that neither Husserl nor Heidegger could fully account for the meaning of this sentence. Certainly, by using their philosophical resources, 'you may give a wonderful account of what you experience with the feeling of hunger', and Heidegger's notion of care, in particular, can be used to describe the way in which food appears as available when I am hungry (*IHJ*: 344). But if I ask *why* I am hungry, and why I need a certain amount of nutrition to satiate the feeling, Husserl and Heidegger's phenomenologies fail to provide satisfactory answers because neither connects the feeling of hunger to the metabolic processes of the body underpinning it. The latter, biological evidence derives from objective study of the body, which can only be understood as pertaining to *me* from the perspective of a fundamentally embodied self.

Jonas's method is able to do precisely this. As indicated, he holds that our corporeal being is, however distantly, involved in every worldly dealing, act of cognition or concern that Heidegger described. This ranges from the use of tools – which can only appear as ready-to-hand 'for a being who possesses hands'

(*MM*: 47) – to the lover's devotion, which is oriented towards the whole person (*PE*: 141). Whomever and whatever I engage with, in 'advancing bodily' through the world I witness in myself 'the point where inwardness actively transcends itself into the outward and continues into it with its actions' (*PL*: 23). Only on this basis, which draws absorption and the body together, can a full account of being-in-the-world be provided.

This is, in short, Jonas's post-Cartesian methodological point of departure. Having set out its basic character we now ought to explain how it contributes to the underlying aim of Jonas's new ontology, namely, a partial rehabilitation of Aristotle's description of nature as immanently teleological. With his phenomenology of embodiment we can say that Jonas has already made a degree of progress in this regard, as 'on the strength of the immediate testimony of our bodies' we know that we move, act and speak with subjective purpose (*PL*: 79). In so describing care as fundamentally corporeal – accounting for our embodied self in a way that neither dualism nor its idealist and materialist offspring could – we can say that on earth there exists, at a bare minimum, *one* physically extended entity that has to be described in teleological terms. And once this has been admitted, materialism cedes its claim to absolute truth. Now, the sceptic might object that as this pertains to human beings alone, immanent teleology cannot be ascribed to other living beings without committing the sin of anthropomorphism. In response Jonas concedes the charge, but argues that in this case anthropomorphism does not, in fact, constitute an intellectual offence. Two arguments are presented in his defence, one hermeneutic–phenomenological and the other ontological.

Jonas firstly notes that the prohibition on teleological descriptions of life lacks sufficient epistemological justification: it is not a *finding* of modern science but a methodological *presupposition* of it, so it should come as no great surprise when, conducted on this basis, biology fails to find evidence of *teloi* in nature. And this proscription is, moreover, a principle that no biologist *qua* human being truly adheres to. The reason is as follows. The biologist takes living beings as their object of study, and proceeds to investigate those beings with a view to deriving greater knowledge of them. Occasionally this involves refining the boundaries of the class of living beings to include some entities and exclude others, such as viruses that appeared to belong to it but in fact do so only partially. All this is true to the scientific method. But on what basis does the biologist initially identify the class of living beings? How do they distinguish the living from the non-living? The answer given cannot be derived from biological understanding, as we would then fall into the trap of circular reasoning.

Jonas's answer is that the life sciences in fact presuppose, as a condition of possibility, an 'almost organic faculty of recognizing life, understanding life, anticipating the encounter of life' deriving from our embodied being (*EDM*: 4). To illustrate this he invites us to engage in a thought experiment. What would a disembodied intellect, such as the mathematician-God of Leibniz, perceive of the living world? The answer, Jonas claims, would be that which Newton discovered in the realm of physics: simply the movement of matter according to efficient causation. Equally, this would be all that we could identify of organisms were it not for the fact that we, as corporeal selves, possess 'peepholes into the inwardness of substance' and thus preconsciously recognize purposeful activity and behaviour (*PL*: 91). To be sure, with this comes the danger of undue anthropomorphism, as in animist ontologies – but this, too, is only a possibility because we have the prior ability to recognize living beings *as* alive. For the biologist as for the philosopher, then, '"teleology" is not a metaphysical afterthought, [...] but a descriptive, phenomenological concept [...] indispensable in progressively organizing the evidence' (*KG*: 163).

Jonas's second, ontological argument against the exclusion of natural teleology emerges from his reading of the Darwinian revolution. In Chapter 1 Darwinism briefly appeared as the last step in materialism's march to ontological dominance, as it seemingly brought humanity, including our faculties and activities, within the purview of mechanistic biological explanation. This is indeed a frequently encountered understanding of its import. But we also intimated that it is, in fact, a misunderstanding.

This interpretation of Darwinism's significance rests on our having accepted the mechanistic account of life as indubitable – otherwise we would have no reason to permit its extension to human beings, given that doing so entails a disavowal of the phenomenological evidence of purposeful behaviour. Yet this immediate evidence is precisely what we are least at liberty to doubt. And once the unity of body and mind in human beings is admitted, our evolutionary connection to the remainder of nature demonstrated by Darwin takes on an entirely new meaning. In a dialectical twist, Jonas suggests that '[i]f man was the relative of animals, then animals were the relative of man and in degrees bearers of that inwardness of which man, the most advanced of their kin, is conscious in himself' (*PL*: 57). On the basis of an integral monism, then, it transpires that the Darwinian revolution, typically thought of as the final victory of mechanistic biology, in fact plays a key part in materialism's overcoming.

Phenomenologically, therefore, we always already distinguish animate being, as purposeful, from the inanimate; ontologically, we have good reason

– in part courtesy of biology – to think that this perception of life is accurate. For both reasons Jonas believes that he is justified in reinterpreting nature on the basis of a phenomenology of embodied existence. From that perspective, which constitutes 'the maximum of concrete ontological completeness known to us', we can survey the remainder of Being, working 'by way of progressive ontological subtraction down to the minimum of bare elementary matter' (*PL*: 23–4).

When we carry out such an investigation, what do we discover? Far from a gnostic dualism between *Dasein* or the *res cogitans* on the one hand and the *res extensa* on the other, we find that there are degrees of care manifest in Being. From our fellow mammals down to microbial life, there is a visible gradation of *freedom*, in perception, feeling, action and thought, running 'like Ariadne's thread' through the living world (*PL*: 3).[5] Such evidence only ends at the rung on the evolutionary ladder where life itself ceases to be. For even in the lowliest unicellular organism, Jonas suggests, a self-concern worthy of the name is discernible in its basic activity. There, in bare life, we find traces of the 'great contradictions which man discovers in himself – freedom and necessity, autonomy and dependence, self and world, relation and isolation, creativity and mortality' (*PL*: xxiii). For this reason the organism as such forms the basis of Jonas's existential interpretation of biology.

III Self-organization

We concluded the previous section by pointing towards the basic activity of the organism as the most rudimentary evidence of self-concern identifiable in Being. But which activity or activities in particular? Genetic mutation? Growth, perhaps, or reproduction? Jonas admits that all are critical for the factual understanding of living beings sought by the biologist. But his analysis, of course, is existential. The focus is therefore on locating the organismic grounds of the absorption that principally characterizes being-in-the-world as care.

Jonas suggests that we find precisely this in the 'teleological structure and behavior of the organism' (*PL*: 91) – features of living beings originally described by Aristotle under the rubric of the 'natural organized body' (1984e: 415b) and its being 'moved by intellect, imagination, purpose, wish, and appetite' (1984c: 700b). As indicated in Chapter 1, both teleological aspects are to be understood as 'immanent' rather than 'transcendent' (*PL*: 34). Here purpose is not imparted by a being that transcends the instantiation of that purpose – as is the case with a tool or machine – but is instead identical *with* that instantiation: the structure of the body is an act of *self*-organization, while the purposeful behaviour of the organism is no less its own than the desire to eat an apple is mine.

Of the organism's self-organization and behaviour Jonas gives priority to the former. The reason why is illustrated by a comparison of three different living beings: a human being who has just had a serious accident, a lizard found by the side of the road and a recently planted sapling. Jonas asks what we would look for in each that indicated the presence of life. In the first case we would search for a heartbeat and establish whether the victim was still breathing, on the grounds that these functions are continuous with the life of the human organism. In the case of the lizard we could, if we were well versed in herpetology, carry out the same tests, but the physical contact that doing so requires would already provide us with an answer in the form of an immediate reaction from it. As such, Jonas suggests that we would probably just 'tickle it with a grass-stalk and see whether it responds' (*OF* II: 2). In the third and final case, we would simply observe over a period of time whether the sapling grew or else withered.

In each case the activity that we hope to find is a different kind of teleological movement. The sapling's growth constitutes formative motion, which adds to the existing structure; the lizard's reaction is an external motion, changing the spatial relation of the body, or a part thereof, to its surroundings; inhalation, exhalation and the beating of the heart are examples of internal motion, which 'seems to change nothing, either with respect to place or to form, but simply coincides with the fact of a body's being alive' (*OF* II: 3). Immediately we realize that, although all three must feature in the life of an organism, the latter is most fundamental since it alone is perpetual: a living being can be fully grown and perfectly at rest, but if it no longer carries out certain internal processes then it is no longer living at all.

The processes in question – to which inhalation, exhalation and the heart's circulation of blood all pertain – are *metabolic*. Metabolism refers to the set of chemical reactions taking place within the organism that serve three different purposes: the anabolic conversion of nutrition into the building blocks of cells, the catabolic breaking down of nutrition and finally the elimination of waste material. Due to both its ubiquity among living beings and necessity for the other kinds of teleological movement described, metabolism, according to Jonas, 'can well serve as the defining property of life' (*MM*: 88).

The most striking aspect of metabolism is that the organism maintains formal continuity throughout the incorporation and excretion of substance, as this implies an 'independence of form with respect to its own matter' (*PL*: 81).

On this basis Jonas suggests that material composition is of merely secondary import to the organism, which in truth *is* self-organization:

> [T]he parts of which the organism consists at a given instant are only temporary, transient contents whose joint material identity does not coincide with the identity of the whole which they enter and leave[,] and which sustains its own identity by the very act of passage of foreign matter through its spatial system, the organic form. This whole is never the same materially and yet it persists as its same self – *by* not remaining the same matter. (*OF* II: 15–16)

Precisely because this turnover of matter is ceaseless, the organism cannot be reduced to its material composition of any given moment. The being of the organism must instead be described in a temporal register, as a 'performance', 'act' or 'process' (*OF* II: 48, 6; emphasis removed). This fact alone is sufficient to mark out the novelty and uniqueness of organismic being within Being as a whole: the organism *is* because it *does*, rather than a being that does things independently of the fact that it is.⁶

The distinction just drawn can be clarified by a comparison of the organism with different kinds of natural and artificial beings. Take, for example, a pebble. Internally it is inert: no process of material exchange occurs within that sustains it in being. Lacking the capacity for self-organization, therefore, the formal identity of the pebble is at any point in time reducible to its material composition. Should it erode from weathering, or split beneath the boot of a passing hiker, then it simply becomes a number of new pebbles: in this eventuality we cannot even say, except metaphorically, that the original pebble is now impaired. For the latter to be true there would have to be a state that the pebble *should* be in, but as it exists in accordance with neither an immanent nor a transcendent *telos*, the pebble lacks the possibility of being either complete or incomplete. This last point distinguishes the pebble from an inanimate artefact. A statue, for instance, is comparable to a pebble in that it lacks the capacity for self-organization: 'in Aristotelian terms, it is not an "entelechy"' (*OF* II: 19). But as the statue is given form through the sculptor's intentional actions, it has an identity that 'is the function of a teleology outside the object itself' (*OF* II: 19). This entails a state that it ought to be in, and for the very same reason it can be accurately described – should the statue fall from its plinth and shatter – as broken.

Now, it might be said that by using immobile beings as our points of comparison, we have stacked the deck in favour of life, as the organism will clearly ontologically differ from a pebble or statue. But what of non-living beings that exhibit movement, in particular those, such as rivers, that are permanently in flux?

To take the latter case, the river is in fact dissimilar to the organism, the statue *and* the pebble. The internal inertia of the last of these, as we have seen, means that left to itself it is both formally and materially stable. The river, by contrast, is in a continual state of material turnover, water both perpetually entering and exiting its boundaries. Should this process happen to stop, then it ceases to be what it is – a river – and becomes something else instead (a lake or a pool). At first this appears not unlike the organism, the processual being of which entails that stasis equates to its death. Yet the comparison with the organism ends here, as the river's turnover of matter does not contribute to any maintenance of form. Regardless of whether the river swells, dries up or flows steadily, none are acts of self-organization, but entirely the result of external factors affecting it. Once again, then, the formal identity of the river is reducible to its material composition at any given moment, a situation only complicated by the fact that the latter changes continually – hence why Heraclitus was right to say that one cannot step into the same river twice.

At this point a critic could concede that the organism is distinct from other kinds of natural being, exemplified by pebbles and rivers, and different again from inanimate artefacts such as statues. But, they might say, this still leaves the Cartesian account of life as mechanism unchallenged. As we saw in Chapter 1, Descartes did not simply describe organisms in mechanistic terms for heuristic purposes, which we might excuse, but went further and suggested that plants and animals actually *are* a kind of machine. Does this interpretation of life – which even today finds adherents among philosophers and scientists – stand up to scrutiny?

Jonas again arrives at an answer through a thought experiment. Suppose, he says, that a person lacking scientific expertise is confronted with a previously unencountered object, and asked whether it belongs to the class of living or non-living beings. What kind of qualities might this person intuitively look for and respond to in the entity in question? Jonas suggests that if, upon touching the unknown being, they felt a body reminiscent of metal, glass or stone, then they would be inclined to classify the entity as non-living. Presumably this judgement would be the result of familiarity, the bodies of living beings tending to be warm and soft and those of machines cold and hard. But this generalization may indicate something important about life itself. Our fictional investigator might try to justify their intuitive response by suggesting that

> a living thing cannot be of steel or glass because these materials are somehow just too rigid and cold; and if pressed further to say why rigidity in the parts

should not be compatible with life if only there is flexibility enough in the joints, he might venture the answer that a material of that kind of rigidity is all too 'set' in its internal makeup to admit of anything further going on inside it once it has been set that way. It is somehow too *definitive* to be alive. (OF II: 9; emphasis added)

Here scientific evidence would vindicate the layman: the properties described are those of non-colloidal substance, whereas living beings, as far as we know, are invariably colloidal. The significance of this fact is that colloidal being is dynamic, and thereby allows internal activity such as metabolic exchange – the defining property of life – to take place. So although not all colloids are alive, all living beings are colloids, and to this extent the layman's intuitive search for the qualities of the latter is appropriate.

Armed with this knowledge we may draw one final comparison. The living being, we said, is a metabolizing being, meaning that it formally sustains itself in being through the act of material reconstitution. While inert artefacts do not, by definition, act in anything like this fashion, we might suppose that complex modern machines do. Take a car, for example: among other things it requires petrol, oil and water to run. Is this not identical to the organism, which equally requires sustenance in order to continue living? In fact the similarity between the two is only superficial, which becomes apparent when we recall that the car utilizes fuel and emits waste as fumes without the fuel ever *becoming* the car. Consisting of non-colloidal substance, it is incapable of doing so, and as such only carries out the two least remarkable of metabolism's three tasks. This is true of all machines, however complex – their fixity entails that they remain materially complete whether switched on or off, using energy or not, and are thus formally reducible, once again, to their material composition. Self-organization, unique to living beings, ontologically distinguishes organisms from machines and so repudiates the purported equivalence Descartes drew between the two.

Having described at length how metabolism sets organisms apart from other kinds of being, a little more needs to be said about what it achieves. In a basic, unicellular organism the anabolic aspect of metabolism is restricted to converting sustenance into the building blocks of the ectoplasm, cell membrane and other simple parts. In advanced organisms, such as ourselves, it does this in the service of reconstituting muscle, fat, skin, organs, nerves and so on. In both cases, however, these parts together form a systematic whole of interdependent functions and processes. The identity of each part, from the cell to the organ, thereby derives from its place in the organism:

> [T]he parts in question cannot exist otherwise than *as* parts of their respective whole (a machine part can): [...] there cannot be a heart, an eye, a nervous system, by itself. Outside the whole these structures lose not only the meaning of their function and the power to function, but the very possibility of existence. Furthermore, they have come into existence through acts of self-articulation of the whole and are thus creations of that which they subserve. [...] [T]herefore, the whole is ontologically prior to this kind of part. (*OF* II: 39)

Jonas's claim that 'there cannot be an eye by itself' is liable to appear strange, and so his reasoning deserves to be spelt out. An organ is not, as its etymology implies, a tool or an instrument, which are extraneous beings that can always be picked up and set down by their user. The organ is instead part of the organism, and part of it in a particular way. An unstructured being – a pile of rubbish, say – simply *is* the sum of its parts, as these could lie in any given way and the entity would remain the same. A house, by contrast, is a structure arranged in such-and-such a fashion, meaning that it is more than the sum of its parts. But still this 'more than' is something *applied* to the parts, which only take on their teleological status through an act of transcendent intent. The organism, however, as a process or act of *self*-organization, is both more than the sum of its parts and – crucially – also generative of them *as* parts. An eye only fulfils its capacity as a means of sight when part of an organism that also sustains its being; removed from this context it loses both its identity and the ground of its being, withering and decaying.

Before turning to the existential significance Jonas attributes to self-organization, we must first address a criticism that his account is open to. It could be argued on the basis of modern genetics that the systemic organization of the organism is in no way its own achievement, but rather the fulfilment of a blueprint encoded in the genome.[7] Moreover, this blueprint is, according to this line of argument, simply the product of evolution rather than anything that the life of the organism bears upon. Thus the organism is, as far as its self-organization is concerned, merely a puppet controlled by extraneous forces. These claims suggest that Jonas's emphasis on metabolism is at least partly misplaced, as, if they are correct, the organization of living beings occurs according to genetic instructions written by the evolutionary process. We shall briefly address both issues.

For the most part Jonas pays little attention to genes, but at one point he does suggest that the genome straightforwardly causes the organism to self-organize. He says that the living being grows 'according to a predetermined "plan"', any deviation from which 'during development is a failure to achieve itself' (*OF* II:

40). The recognition of possible deviation entails that Jonas does not endorse a naive genetic determinism, which is fortunate, as genetic sequences are in fact only 'indeterminate resources' of development (Moss 2005: 359). Nevertheless, his reference to deviation as a 'failure to achieve itself' implies that the organism is *supposed to* conform to a genotypic blueprint. But this, too, is inaccurate:

> The empirical fruits of several decades of research in molecular, cell, and developmental biology have revealed that what distinguishes one biological form from another is seldom, if ever, the presence or absence of a certain genetic template but rather *when* and *where* genes are expressed, *how* they are modified, and into *what* structural and dynamic relationships their 'products' become embedded. (Moss 2003: xvii)

The host of environmental and circumstantial factors involved in development shifts the focus back, at least in part, towards the life of the organism as critical for self-organization. With this move, the idea that a genotypic 'plan' serves – or should serve – as a straightforward explanation of organismic development is undercut.

The second issue was whether the genome is the product of evolution alone. Once more, Jonas appears on one occasion to subscribe to this idea: 'Improvement of the plan itself, new adaptations, can come about only in the succession of new organic individuals – each by itself as determined in its particular plan as every other individual organism' (*OF* II: 40). But again, this has since been shown to be inaccurate: organisms are not, in fact, passive bystanders in evolutionary history, for the following reason. Not only is phenotypic development, up to and including physiological functioning, informed by environmental factors but, crucially, the organism can also make changes *to* its environment – a process known as 'niche construction' – which then feed back *into* the development of the organism (Barker 2015: 62–6). The upshot is that even over the course of a lifetime a living being can alter its own development, before informing the evolutionary process through reproduction. The role of the genome in phenotypic development and evolution is therefore more nuanced than Jonas realizes, which only vindicates his decision to emphasize self-organization through metabolism instead.

IV Behaviour

We stated earlier that self-organization is just one of the teleological aspects of the organism. The second is behaviour, which here refers to acts either voluntarily or involuntarily undertaken in relation to oneself or one's surroundings. Obvious examples are the acquisition of food, the pursuit of sexual reproduction and the avoidance of predators. But Jonas also has in mind the basic capacity of an organism to orient itself in its environment by relating to beings therein. In Chapter 3 we will focus on the different ways in which plants, animals and humans behave, bearing out Jonas's claim that gradations of care are manifest in Being; for now we shall look at the phenomenon in its broadest sense, so as to understand why Jonas believes that care is co-existent with life.

Returning to our earlier example of external motion – tickling the lizard with a grass-stalk – we recall that Jonas suggests carrying this experiment out in order to ascertain whether the lizard is alive. Implicit in doing so is not merely the anticipation of *movement*, but of a *response*. The difference between the two goes to the heart of the distinction between living and non-living being. When we push at a door we expect it to swing away from us as a result, but only in a mechanistic fashion explicable by efficient causation and a transcendent *telos*. When we tickle the lizard, however, we bring with us the knowledge of our own embodied being in anticipating a reaction guided by internal purposes. Let us elaborate on this latter idea.

In our encounter with the lizard, we are sure we are dealing with a being that – provided it is not dead – responds to the touch of the grass-stalk according to certain instincts or urges. Why? Because we know from our preconscious understanding of life that this capacity 'interposes between the affection from without and the reaction from within' (*OF* II: 3). The latter refers to purposes either conscious and deliberate or preconscious and instinctual, which we see in the movements and comportment of other organisms, and bear witness to in ourselves. Thus it is not a mechanistic response: 'For the goad urges the ox to move not by imparting its momentum to his body but by calling on his own urges or drives: only in thus being met at the receiving end by a self-concern can the external incidence, and even its being sensed, become a "stimulus"' (*OF* II: 4). This explains the tentativeness with which we disturb the lizard: Will it fight or flee? Here is a demonstration of the degree to which the anticipation of life features in our experience, searching for an appropriate object. This was surely true even for Descartes, who would not have endeavoured to explain away the impression of animal subjectivity unless he had shared it.

But – the sceptic will respond – how do we *know* the lizard is an appropriate object, given that all we can measure of its behaviour is the external aspect? There are, after all, machines that externally act in a similar way. Robots are the most obvious example, but even common cybernetic devices such as thermostats are, through feedback loops, able to take cues from the environment and then regulate their own functioning so as to modify that selfsame environment. Does this not suggest that both beings are equivalent in their activity, even if different in their constitution?

Jonas believes not, arguing that both 'teleology, i.e., having a "sake", and awareness, i.e., being internally affected by the environment, emerge only and entirely *with* that organization which we call organic' (*OF* II: 33). His reasoning as to why is as follows. In attributing interiority and sensitivity to a being we suppose three things: firstly, that it is *a* being, a single entity; secondly, that its sensitivity is spread (albeit unevenly) across its unified form; and thirdly, that its purposes serve the whole organism. In all three ways we implicitly regard the living being as an individual. But what is the condition of possibility for this singular self? The answer can only be metabolic self-organization. Of the forms of organization analysed earlier, the being of the organism alone was one of material reconstitution in pursuit of an enduring formal identity. The stone and river, as discussed, lack any ability to self-organize and are thereby entirely reducible to their material composition. The machine, by contrast, does have a formal identity, but it is still not one of its own doing: the being of the machine – even one, such as the thermostat, that takes cues from the environment – is not processual but rather an arrangement of otherwise inert matter. Whatever is of relevance to the cybernetic device in the external world is merely appropriated as an input into a fixed and self-sufficient system, which would endure whether switched on or off. And the reason for this harks back to the basic nature of a machine: that its teleological principle is not immanent but rather transcendent. Sensitivity and interiority can therefore be attributed to living beings alone, meaning that purposeful behaviour appears 'in its most elementary form' in the 'irritability' of the single-celled organism (*OF* II: 3).[8]

Having discussed behaviour and metabolism at length, it now falls to us to bring out the full existential significance of this twofold teleology (again, with the caveat that Jonas himself does not always explicitly connect his account back to Heidegger's). We may firstly observe that metabolism allows for the possibility of being-in-the-world, initially via a more rudimentary being-in-the-environment. Why? Because in self-organization the organism wrests itself from the remainder of Being. In this act of individuation the living being creates the boundary that demarcates itself as an entity amid other entities. As Jonas puts it: '[O]pening into an environment […] is grounded in the fundamental transcendence of organic form relative to its matter' (*PL*: 84). Now, the metabolic creation of the self-world/environment distinction is by itself insufficient to account for *openness* to beings within the latter, which is what it means to 'have' an environment. How, then, does Jonas explain this? The answer lies in our discussion of sensitivity. As discussed, in the act of self-organization the organism creates a boundary between itself and the world. Regardless of whether the boundary in question is the cell wall of an amoeba or the skin of a human being, it has sensitivity or irritability. As such, from the organism's inside 'the rest of reality is the "other", the external: crowding in upon it from out of the environment' (*OF* II: 46). These other beings meet with and so stimulate the sensitive or irritable boundary of the organism, alerting it to their presence. This is, Jonas suggests, the corporeal root of being-in-the-world that Heidegger had described from the point of view of *Dasein*. In Chapter 3 we shall see how Jonas explains the transformation of the organism's environment into a world proper, but for now we may note that he is able to explain something Heidegger could not, namely, the origin of the former.

We recall that, crucially for Heidegger, being-in-the-world initially and for the most part takes the form of instrumental 'dealings' with beings. For Jonas, having explained openness to the environment as a corollary of metabolism, dealings are therefore necessarily an accompaniment of metabolic being. Precisely because its material composition is perpetually changing, the organism is committed to acquiring adequate nutrition so as to continue in being, and its sensitive boundary then allows for engagement with the environment that contains this sustenance. He tells us that in 'single encounters this otherness has the quality of foreign body or influence which is either useful or harmful; in its entirety and as an enveloping horizon it has the character of "the external world" confronting the organism's overwhelming concern in its own life-process' (*OF* II: 46). Just as *Dasein* encounters the majority of beings as available, therefore, the organism is initially and primarily engaged with beings insofar as these relate to its needs. Of course, Jonas does not conceive of dealing-with and openness to the environment as equally realized across the organismic realm in terms of richness or intensity. His point is rather that the organismic encounter with beings as something-for-nutrition is 'the fundamental condition from which ultimately all "later" characteristics of life derive' – a progressive achievement that we shall look at in Chapter 3 (*OF* II: 47).

As with *Dasein*, therefore, the organism is fundamentally absorbed in its environment. But what are the dimensions of this engagement? Jonas's answer, again following Heidegger, is space and time: each representing a horizon 'into which life continually transcends itself' (*OF* III: 2). We shall take these in turn.

Heidegger had defined space not as the objectively measurable grid of physical science, but rather primarily as an aspect of absorption. *Dasein* is always in the process of 'de-distancing', he claimed, which is not to be understood as saying that engagement with an entity necessarily brings it measurably closer to us (2010a: 102). He instead meant that a being which shows up in our dealings is always nearer to us than one that lies unnoticed, however physically close the latter may be. 'When we walk', Heidegger says, 'we feel it with every step and it seems to be what is nearest and most real [...]. And yet it is more remote than the acquaintance one meets while walking in the "distance", twenty steps away' (2010a: 104). This is all phenomenologically sound, and the reference to the embodied act of walking is welcome. Alas, Heidegger went on to equate corporeality itself with 'the I-thing encumbered with a body', which is spatial, he thought, only in the sense of being mathematically calculable (2010a: 104). Certainly, the objective body of scientific understanding pertains to spatiality in this sense alone. But what Heidegger again misses is the lived body, which is fundamental to any adequate analysis of phenomenological spatiality, as Jonas demonstrates. We mentioned earlier that being-in-the-world originated, for Jonas, in the organism being 'directed towards the co-present *not-itself*, i.e., the "environment" – that total "other" [...] which holds the stuff relevant to its own continuation' (*OF* III: 2). This 'directedness constitutes biological *space*' (*OF* III: 2): the horizon in which beings have the character of 'near' or 'far' as they relate to the lived body and its absorption grounded in need.

In much the same way, Jonas argues that the temporality of *Dasein* has its origins in organismic being. As discussed, temporality was paramount for Heidegger's existential analytic. Care, the being of *Dasein*, consists of projections into the future informed by the past, entailing that temporality is its 'ontological meaning' (2010a: 309). But Heidegger sought neither to explain where this meaning ontically came from, nor whether non-human life had a share in it.

Jonas agrees with Heidegger insofar as he takes temporality to be the ontological meaning of care, but of course differs in holding that the latter is grounded in the organism's physical needs: '[S]elf-concern actuated by want throws open [...] a horizon of time embracing, not outer presence, but inner imminence' (*OF* II: 62). Equally, like Heidegger Jonas takes temporality to principally consist of projections into the future, rather than the deliverance of the past. But again, he locates the basis of this phenomenological observation in organismic being. Metabolism, the process that gives rise to absorption, entails that the organism is uniquely in a state of becoming: 'While mere externality is, or can be presented as, wholly determined by what it was, life is essentially also what it is going to be' (*OF* II: 63). For this reason 'the extensive order of past and future is intensively reversed', the latter taking precedence over the former

(*OF* II: 63). Of course, life that bears minimal sentience is unlikely to *experience* time in anything like the same fashion as self-conscious human beings, and Jonas sensibly makes no such claim. His point is simply that metabolic being opens a horizon of transcendence, one that only later – perhaps only in humanity – blossoms into that which we experience as the flow of the present unfolding into the future.

The last aspect of Heidegger's existential analytic that Jonas directly grounds in the organismic constitution is being-toward-death. Understood as the possibility of *Dasein*'s impossibility, the recognition of finitude was, for Heidegger, both a uniquely human and distinctly intellectual affair. But death is only a possibility for *living*, corporeal beings. Where in Heidegger's account of being-toward-death, then, were debilitation and decay, the cry of the flesh and fear of spilt blood? The closest he came to accommodating death's visceral nature was in his 1929 lecture course. Briefly addressing 'the motile character of the living being as such', Heidegger entertained the possibility that an animal's flight from danger or shrinking in pain meant that they had some kind of awareness of death (1995: 266). Yet because he did not understand *Dasein*'s being as embodied, Heidegger could not connect our being-toward-death to the animal's fear of dying. As such, he could only concede that 'the natural physiological death with which the particular living individual intrinsically comes to die […] represents a central problem' (1995: 267).

For Jonas, by contrast, being-toward-death 'resides in the organic constitution as such, in its very mode of being' (*MM*: 88). The reason why harks back to the peculiar, even paradoxical nature of metabolism. We noted that in its processual being the organism had achieved a freedom of form from matter: for as long as it lives it is irreducible to its material composition.[9] This freedom is by no means unconstrained, however, but on the contrary a '*needful freedom*' (*PL*: 80). Precisely because the organism lives in maintaining a formal identity through the perpetual turnover of matter, it *must* succeed in obtaining adequate nutrition if it is to continue being. Put epigrammatically: the organism is what it does, and so it must do in order to be. Crucially, this necessity brings with it the ever-present possibility of the organism's demise: the failure to acquire or incorporate sustenance brings the process of its being to an end. Thus, conscious of the fact or not, the life of the organism 'has in it the sting of death that perpetually lies in wait, ever again to be staved off' until, one way or the other, death claims it at last (*MM*: 91).

From the preceding discussion we can see, then, that being-in-the world (initially the environment) first manifests as purposive behaviour, which is itself a corollary of metabolism, entailing that the latter is the former's condition of possibility. The behaviour of the organism, its dealings with beings, has the dimensions of spatiality and temporality, and ultimately takes place against the backdrop of being-toward-death. These are the components of the care-structure – *Dasein*'s being-ahead-of-itself-being-in-the-world – that Jonas traces back to life as such, and are implied in his straightforward description organisms as beings 'whose being is committed to their own care' (*OF* II: 31).

V The *nisus* of Being

With this last point we conclude our reconstruction of Jonas's existential analytic of the organism. Yet one question remains to be posed: namely, why did life come to be in the first place? This is, of course, one of the biggest questions of all, and so we cannot hope to arrive at any kind of definitive answer. Nevertheless, Jonas offers one, and as we shall see, it later acts as the basis for one of his key axiological claims.

Insofar as the question of life's origins can be answered, Jonas holds that biology currently lacks the resources to do so. Evolution by natural selection is, no doubt, the best causal explanation we have as to why certain forms of life give rise to other, more complex forms. This it does by pointing towards the increased chances of survival afforded by environmentally beneficial phenotypic traits, the genetic mutations associated with which are then more likely to be passed on through reproduction. But it cannot explain why organisms – that is, the kind of entities that are subject to the pressures of natural selection – came to be in the first place. For how could the criterion of greater environmental fitness apply to non-living beings, which are themselves under no such pressures to thrive and reproduce? As Jonas says, '[T]he survival standard itself is inadequate for the evolution of life. If mere assurance of permanence were the point that mattered, life should not have started out in the first place' (*PL*: 106).

Biological science may yet, to be sure, find a solution to this problem – at the very least, the possibility cannot be ruled out. But in the absence of one only reasoned speculation can satiate our philosophical curiosity, and Jonas believes that the method underlying the preceding analysis – an existential-phenomenological interpretation of biological facts – may have something to offer in this regard. Jonas first suggests this on the opening page of *The Phenomenon of Life*. There he says that 'since matter gave such account of itself, namely, did in fact organize itself in this manner and with these results, it ought to be given its due' (*PL*: 1). But what does giving matter its due in light of its results entail? Clearly, the answer cannot – if it is to be philosophical rather than theological – invoke a divine agent as responsible for the emergence of life from non-living Being. This would involve, on Jonas's schema, a transcendent *telos* underpinning the workings of nature, which nothing in the testimony of life described earlier invites us to posit. We are searching, instead, for a cause that remains 'entirely within the immanent', and thereby consistent with a monistic, anti-gnostic metaphysics (*MM*: 173).

Jonas arrives at a tentative answer through a critical engagement with the panpsychism of Alfred North Whitehead. Whitehead's 'philosophy of organism' was undoubtedly a significant influence on Jonas in the post-War period (*M*: 195); indeed, he credits Whitehead with an 'intellectual force and philosophical importance [...] unequalled in our time' (*PL*: 96). However, Jonas rejects Whitehead's suggestion that some form of interiority could be ascribed to all entities, from human beings down to the level of elemental matter. Whitehead had in this way sought to wholly overcome Cartesian dualism, fully identifying mind, or degrees thereof, with material substance. Thus in Whitehead's ontology the problem of the emergence of life from matter was not so much solved as dissolved: immanent teleology – which we have suggested characterizes organisms alone – belongs, on this schema, to all physically extended beings, living or not.

Jonas objects to this as being 'overbold' (*MM*: 211) and 'uncalled for by the record of reality' (*IHJ*: 357). For one thing, when we bring our pre-theoretical understanding of embodied being to bear on inert matter, we find neither a sensitivity that would betray purposive behaviour, nor the capacity for self-organization that would underpin such concern. If we handle a piece of glass – whether fashioned by humans or the forces of nature – it offers no reaction or even irritability, merely a passive resistance to pressure exerted upon it. Equally, the formal persistence in being of the glass is not the result of any self-organization on its part, but instead depends entirely on the external forces that come to bear on it. All this holds not only at the macroscopic level of the single piece of glass but also at the microscopic level of its silica composition.

In his commitment to panpsychism Whitehead was also led to overlook the unique self-concern of living beings. Precisely because, according to Whitehead, non-living Being has its share of sentience, there is 'no real place for death' in his ontology: for the organism to die is simply for one sentient being to become many smaller sentient beings (*PL*: 95). Yet as we have seen, the teleological activity of the organism is oriented towards continuation in being and away from non-being, the latter explaining the 'deep anxiety of biological existence', which is similarly absent from Whitehead's system (*PL*: 96). In both death itself and the being-toward-death of the organism we bear witness to aspects of care that are absent, as far as we know, from non-living beings. For Jonas, therefore, Whitehead responded to the problem of the emergence of life in Being only by stripping life of its most conspicuous dimensions, at the same time as attributing too much of the latter's teleological nature to non-living Being.

Jonas therefore opts for the more parsimonious claim, which nevertheless remains consistent with his analysis of life. The care-structure that characterizes the organism is, he suggests, a *potential* of non-living Being: it is there in a dormant state, remaining so unless and until the conditions that permit its stirring happen to arise. In Jonas's words, '[R]ight from the beginning matter is subjectivity in its latent form, even if aeons, plus exceptional luck are required for the actualizing of this potential' (*MM*: 173). This postulation represents, then, the deepest point of Jonas's attempt to overcome gnostic dualism with a monism that nevertheless does justice to the uniqueness of life.

Now, although this conjecture is an elegant attempt to reconcile the continuities with the differences that hold between living and non-living Being, it still does not solve the puzzle of why non-living Being should have given rise to life at all. The reason for this is that, unable to identify self-organization and purposive behaviour in matter, the latter's development into life cannot be attributed to an immanent *telos* in the sense in which this notion has been employed so far. What, then, does reasonable speculation, based on matter's latent capacity for care, permit?

Again Jonas opts for the more parsimonious claim. He suggests that we would have to ascribe to matter not a *telos*, but rather 'a tendency, something like a yearning' towards the realization of life (*MM*: 173). This he calls a 'cosmogonic eros' towards life (*MM*: 166), in reference to the *Lebensphilosoph* Ludwig Klages, who had sought to explain the original emergence of life with this very concept. The invocation of *eros* carries connotations of emotion and desire, however, and – as we shall see in Chapter 3 – these are capacities of animals and humans alone. For this reason it might be less misleading to speak of a *nisus*, which suggests not a striven-for end, exactly, but merely a tendency in one direction.[10] Following Jonas, then, we might speculate that a *nisus* corresponding to matter's latent self-concern was the non-mechanical cause that brought about the original development of life.

The *nisus* of Being, this primordial tendency in matter towards the living, has as far as we know met with success on our planet alone. Despite this,

Jonas suggests that '[i]n organic life, nature has made its interest manifest and progressively satisfies it, at the rising cost of concomitant frustration and extinction, in the staggering variety of life's forms' (*IR*: 81). What interest, exactly? That of world-openness, of degrees of *freedom*, realized in the development of powers of motility, emotion and thought. This seems, at first, to contradict our earlier statement that evolution by natural selection is the best theory we have to explain the emergence of higher forms of life. Yet they may in fact be compatible, the former explaining what it is in life that the *nisus* of Being is oriented towards, and the latter accounting for the process by which this has – blindly, no doubt – come to be. Although Jonas admits to being very much on the terrain of speculative metaphysics at this point – which may prove problematic, given the weight that his axiology and ethics will place on the *nisus* of Being – he wonders whether 'in the striving of total reality it was the meaning of the "experiment" to redeem things from the muteness of their self-enclosedness […]. At any rate, having a world must be regarded as part of the original venture of life, [and] its different modes therefore as modes of life's fulfilment' (*OF* V: 66–7).

With his existential analytic of the organism Jonas sets about turning the page on modern dualism and its monistic progeny. Rather than rejecting the Cartesian bifurcation of nature into mind and matter only to exclusively accept the latter – as many contemporary philosophers either intentionally or unwittingly do – Jonas weaves the two threads back together. The result is a phenomenological ontology that respects the scientific facts at the same time as disputing their materialist explanatory framework, the latter being ultimately incompatible with the efficacy of consciousness. Only an integral monism of this sort can account for the full record of Being, in which life must take pride of place. Regardless of whether Jonas is correct in his speculations as to the origins of life, we may say that its development, some four billion years ago, set the stage for an evolutionary adventure through the ascending levels of plants and animals, before culminating in human beings. This increasing world-openness that the biological record presents us with – the scale of nature, or *scala naturae* – will occupy us in Chapter 3, and complete our reconstruction of Jonas's philosophy of life.

3

The philosophy of life II
The *scala naturae*

I Aristotle after Darwin

In the preceding chapter we reconstructed the core of Jonas's existential interpretation of biological facts. Jonas neatly summarizes this account in the following maxims: '[T]here is no organism without teleology; there is no teleology without inwardness; and [...] life can only be known by life' (*PL*: 91). To a great extent this amounts to a rehabilitation of Aristotle's conception of the organism as immanently teleological. In the present chapter we shall see that Aristotle also plays a key role in Jonas's analysis of the three broad classes around which life's manifold forms coalesce: the vegetative, the animal and the human.

The conception of the living world as gradated according to these divisions is traditionally known as the *scala naturae*, or the great chain of Being. This notion can be traced back, once again, to the work of Aristotle, as Jonas notes:

> Aristotle read this hierarchy in the given record of the organic realm with no resort to evolution, and his *De anima* is the first treatise in philosophical biology. The terms on which his august example may be resumed in our time will be different from his, but the idea of stratification, of the progressive superposition of levels, with the dependence of each higher on the lower, the retention of all the lower in the higher, will still be found indispensable. (*PL*: 2)

The notion of stratification that Jonas favourably cites refers to Aristotle's doctrine of the soul (although, as discussed in Chapter 1, what Aristotle meant by '*psychē*' is better captured by 'principle of life' than 'soul' as it is used in modern English). He suggests that the soul has five accumulative parts: 'the nutritive, the appetitive, the sensory, the locomotive, and the power of thinking' (1984a: 414e). The first two parts, relating to growth and reproduction, constitute the vegetative soul possessed by plants; the next two parts, sensation and locomotion, constitute the sensitive soul that animals have in addition to the first pairing. Humans alone, however, are endowed with all four *and* the capacity to reason, hence our status as the *animal rationale*. As we shall see, this is a division of capacities that Jonas broadly adheres to.

The qualification in Jonas's praise is due to the fact that Aristotle, unaware of evolution, conceived of species as embodying unchanging essences. In the *Generation of Animals*, for instance, Aristotle tells us that it is impossible for an animal 'to be eternal as an individual [...] but it is possible for it as a species' (1984a: 731b). This neatly illustrates Aristotle's general conception of nature as stable and enduring, itself exemplary of the broader Classical world view, 'which took the cosmos to be a harmonious system' (*DPL*: 117). Within such a system there is no room for a 'history of the world or a history of nature, nor even a history of parts of nature' – here 'becoming' is a principle of individual beings alone, rather than the 'inner character of Being' as such (*DPL*: 117).

After Darwin and the expansion of the fossil record, however, we know firstly that Aristotle's belief in the eternal essences of species is false, and secondly – as a direct consequence – that the Classical notion of a harmonious, ahistorical cosmos is mistaken. On the contrary, history defines nature through and through, both in the rise and fall of species within their ecosystems and in the changes that species and ecosystems themselves undergo. Most radically of all, of course, we know ourselves as humans to be subject to the very same processes. In contrast to the ancients, therefore, our belief can only be in 'becoming rather than abiding nature' (*PL*: 283).

Nevertheless, Jonas argues that the discrediting of the Classical version of the *scala naturae* does not mean that the notion is nonsensical per se. On the contrary, he claims that Aristotle's analysis of the soul glimpsed a fundamental truth, one entirely consistent with a post-Darwinian philosophy of becoming. As indicated in Chapter 2, Jonas argues that life, in evolving, gradually realized aspects of 'world-openness' (*DPL*: 125). This is empirically identifiable by the presence of 'greater sophistications of form, the lure of sense and the spur of desire, the command of limb and powers to act', and finally, in humanity, 'the reflection of consciousness and the reach for truth' (*PL*: 2). With this scale of organic achievement we may identify 'higher' and 'lower' life forms corresponding to the levels of plants, animals and humans, exactly as Aristotle had described.[1]

Such clean-cut divisions might seem at odds with Jonas's broader monism, which, as we have seen, identifies care throughout the living world. If so, it should be reiterated that Jonas only holds that (part of) the care-structure is *formally* present in all organisms, while dramatically differing in terms of the intensity and richness of its manifestations. This is true not only of the variation between the classes of plants, animals and humans but also within these classes, as indicated by the vast existential distance between bivalves and the great apes. What Jonas attempts to identify, however, are the key empirical indications of existential advance, which do indeed coincide roughly with the threefold scale outlined.

In reconstructing his account of the *scala naturae* it will become apparent that certain details of animal and plant life are absent from Jonas's analysis. For the most part this is simply due to discoveries having been made in the decades since he wrote *Organism and Freedom* and *The Phenomenon of Life*; occasionally it is the result of Jonas misreading biological facts that were available to him. We shall see, however, that these facts can usually be incorporated into his theory in a way that complements, rather than compromises, his general account of organismic being. We shall begin with his analysis of plants.

II Plants

As discussed in Chapter 2, Jonas holds that the being of the organism per se is its absorption in beings encountered in its environment. This can be identified in its most rudimentary form in the irritability of microorganisms, such as amoeba. When we turn our attention to plants, however, we find that the phenomena of self-organization and purposive behaviour – the conditions of possibility for care – reach new heights of complexity. Here the 'germ of sensing unfolds to [a] distinct world-relationship, just as the cells grow into the differentiated, composite organism' (*PL*: 99). Let us unpack precisely how this occurs, and what it tells us about vegetative being.

As an organism the plant is of course a metabolizing being, meaning that it self-organizes through the process of material exchange. Equally, the form it maintains in being is systemic, each part carrying out a function that contributes to the life of the whole. Most obviously, the plant attains through its leaves and roots the sunlight, water and minerals necessary for it to metabolize. Since the sources of this sustenance – the sky and earth – are easily located on a vertical axis, the greatest demand on the seed-form of the plant is taking root in a suitable location, should fortune happen to deliver one to it. Jonas notes that this is the 'most efficient means of exploiting the inherent advantages' of photosynthesis as a form of metabolism: through root and leaf 'we have immediacy guaranteed by constant contiguity between the organs of intake and the external supply' (*PL*: 103).

However, the plant's fixity of position and ready satisfaction of needs means that its being remains comparably basic. 'With its adjacent surroundings the plant forms one permanent context into which it is fully integrated, as the animal can never be in its environment,' Jonas claims (*PL*: 104). He goes on to suggest that, as a result of this 'non-motile existence' (*OF* IV: 62), the life of the plant 'is passive in relation to space and active only in its internal processes. External things have to happen to it or to refrain from happening: suitable matter must come near, actual contact has to come about by the accidents of environment so that the freedom of metabolism can come into play' (*OF* III: 3). Though the plant is open to its environment, therefore, Jonas suggests that the relation of care that it enjoys is restricted to a 'dimension of dependence and necessity', rather than the full dimension of freedom that motility allows for (*OF* III: 4).

This is the core of Jonas's account of plant being as it appears in *The Phenomenon of Life*, and it successfully achieves an existentialist recovery of Aristotle's account of the vegetative soul. It is factually incorrect on one major point, however, which undermines the analysis there given. Although plants lack *locomotion* – the ability to move from place to place – they quite clearly possess *motility*: a flower follows the movements of the sun, a tree's roots wrap around rock and stone to anchor itself in the earth and a Venus flytrap snaps shut to catch its prey. All are striking examples not of formative movement, particular to growth – which the plant also undergoes – but of the 'external' kind of movement that demonstrates sensitivity to certain stimuli in the environment. Presumably Jonas failed to acknowledge this due to the pace at which it occurs, as, aside from that of the Venus flytrap, plant motion is generally so much slower than our own that it is all but invisible to the naked eye. Whatever the reason, Jonas apparently recognized his error and corrected himself in the slightly later essay 'Biological Foundations of Individuality'. There he accepts that plants are motile beings, and even concedes that 'for the opening and closing of blossoms and even for the startling performance of certain insectivorous plants […] the outward likeness to the animal pattern is indeed strong' (*PE*: 205 n.9). He maintains, however, that it is of a qualitatively different sort in several respects.

First of all, vegetative movement is typically highly restricted in speed and scope, outstripped in this regard even by the leisurely sloth. This much is obvious. Crucially, he also suggests that it is 'continual', 'predefined' and 'irreversible', consisting of rhythmical stimulation rather than moment-to-moment reorientation (*PE*: 203).

> The turning of the [plant's] leaves is, as to occurrence, rate, and rhythm, entirely governed from without; and it is only owing to the cyclical nature of the light-changes involved, that is, to the external agent, that the phototropic motion does 'revert'. It is not the motion in progress that has the capacity to revert itself; there is but one direction open to it at any given phase. And its change of direction is as reactive as is the bare fact of a motion as such. (*PE*: 205 n.9)

With this additional observation Jonas's analysis is far closer to the truth. Even if we film the plant and play the footage back at a faster rate, so as to bring its irreversible movements within our perceptual range, our impression remains one of the plant being led by other beings rather than acting autonomously in relation to them.

Underpinning these differences in movement is the key distinction between the systemic forms of animals and plants. Jonas suggests that a lack of 'central control' is at the bottom of the plant's world-poverty (*PE*: 205 n.9). Evidently he does not have in mind the capacity for *conscious* control of movement, given that according to his basic analysis of the organism this is not necessary for immanent teleology. He is instead referring to the centralized nervous system, which 'raises the unity and individuality of the organism to an entirely new level' (*PE*: 203). Lacking this centralization, the plant is unable to control its external movement in a coordinated fashion: although leaf, root and stem all contribute to the life of the whole, they move quite independently of one another. Indeed, Jonas goes even further and suggests that the '*morphological* individuality of a tree is not matched by internal individuality evident in *centralization*, such as distinguishes even the solitary amoeba' (*OF* IV: 9). Jonas does not, to be clear, suggest on this basis that a tree has an environment that is *less* rich than that of an amoeba. But he does stress that the plant, despite its clear abilities to engage with certain beings in its environment, is existentially far closer to the amoeba than it is to the animal.

Although nothing in Jonas's later analysis of vegetative being is necessarily incorrect, it still strikes us as somewhat inadequate. Surely, we feel, the being of a plant is richer than that of a single-celled organism. That this is indeed the case can be demonstrated with reference to the field of plant 'intelligence' that has emerged in the last decade. More specifically, there is accumulating evidence for the ability of plants to communicate with one another: through the release of volatile organic compounds in the air and soil it seems they are able to alert neighbouring plants of danger, allowing the latter to employ any chemical defences that might deter herbivorous animals (Karban 2015; Blande and Glinwood 2016). Intriguingly, other studies indicate that such communication is greater between kin, suggesting that it might serve to evolutionarily benefit the reproductive group (File et al. 2012; Karban et al. 2013).

The significance of these findings for an existential interpretation of life is as follows. Jonas claims, we recall, that various dimensions of the care-structure are identifiable in all organismic being: temporality and spatiality, dealings, environment and being-toward-death. All are implied in his description of plant being, although he thinks that these structures manifest scarcely more intensely there than they do in single-celled organisms. Yet even if this latter judgement is correct, the aforementioned evidence allows us to say that plants have a richer being than unicellular organisms: unlike the latter, they demonstrate a dim trace of the mode of absorption Heidegger had called concern (*Fürsorgen*). As before, it seems that evidence of it goes far deeper into the living world than Heidegger acknowledged. In sending warning signs out to its kin alone, the plant relates to the set of beings most closely like itself in a manner distinguishable from the instrumentalism of useful dealings: not its own being is defended, but that of the reproductive group it belongs to.

It feels odd, no doubt, to think of plants as social beings, even to a minimal degree. From Aristotle down to the early modern era, Western thought has attributed to plants the capacities of the nutritive soul alone, a history that continues to exert a powerful effect on the collective imagination. As we have seen, Jonas's philosophy of life also adheres to this view. But – assuming the emerging evidence of plant communication is correct – our hesitancy to break with this account by crediting plants with more (rather than less) of the care-structure would be merely a prejudice, the overcoming of which bears out our intuition that plant life is richer than Jonas recognized.

III Animals

Jonas suggests that with the evolutionary arrival of animals the cosmos bore witness to a significant existential advance. Following Aristotle's account of the animal's sensitive soul, his existential description emphasizes their capacities for 'perception, emotion, and movement' as 'three modes' of greater freedom (*PE*: 206). Jonas holds that together these features signify the realization of a 'real-world relation', in contrast to plants, which are poor in world since possessing only an environment (*OF* III: 1). Jonas's analysis of animal being – occupying most of the third and fourth chapters of *Organism and Freedom* – combines a rich phenomenology of the senses and action with a reading of the biological record. Although we cannot hope to capture the nuances of his account here, we shall attempt to do justice to its key features.

Our entry point into an existential interpretation of the animal form of life again relies on phenomenological description. For what alerts us to the presence of an animal when surveying a particular scene is not immanently teleological movement as such – which, as discussed, plants are also capable of – but rather the noticeable *speed* at which it moves. To be sure, we noted earlier that the external motions of certain plants, such as the touch-sensitive Venus flytrap and *Mimosa pudica*, are comparable in this regard to animal movement. But these are rare exceptions: as a general rule plants move at a pace imperceptible to us, whereas the external motions of animals are typically visible. Of course, the 'sophisticated or scientific observer is apt to smile at the naiveté of a distinction by such relative and "accidental" standards as slowness and quickness relative to a particular extraneous observer' (*OF* IV: 16). Jonas contends, however, that the intuitive observation contains its own wisdom. The two sorts of external motion, perceptible and imperceptible, 'represent really different biological orders of event, of which the one is by its own nature related to perception and the other by its own nature is not' (*OF* IV: 18).

Jonas justifies this claim by pointing to the very different demands of vegetative and animal metabolism. As noted, plants typically convert readily available inorganic substance – sunlight, nutrients, water – into organic matter. Animals, however, are nourished by water and the organic. This basic distinction has positive and negative consequences for both forms of life. In one sense the plant is superior to the animal, as it enjoys the 'amazing power to build itself up directly from the ever-ready mineral supplies of the soil, where[as] the animal has to depend on the presence of highly specific and non-permanent (because highly corruptible) organic compounds' (*OF* IV: 49). In another sense, however – and paradoxical as it may sound – the greater demands placed on the animal by its metabolism are precisely the cause of its different kind of movement, indeed, of its richer being. Jonas argues that animals move at the speed they do because locomotion is essential to the acquisition of organic sustenance. The object of their satisfaction lies at a spatial and temporal distance, and so nutrition must be sought out, rather than made available simply by being there, as in the case of the plant. This ambiguous difference extends even to their otherwise common dependence on water: while the plant can draw the most meagre traces of moisture from the earth, the animal has to actively find a freshwater body – stream, pool, puddle or droplet – and the same is of course true of the animal's broader nutritive requirements, whether herbivorous, carnivorous or omnivorous in kind.

Essential to carrying out this task is the greater range of bodily movement permitted by the animal's particular form of self-organization. We saw earlier that Jonas takes the plant's relative lack of integration to be the cause of its existential poverty; conversely, he suggests that the centralized nervous system is at the root of the animal's genuine world-openness. In the first instance it allows, Jonas suggests, for movement that is variable and reversible, fast and spontaneous. In stark contrast to the flower that follows the gradual arc of the sun, we witness in, say, a beetle the free ability to crawl forth, pause, retract a limb, and to turn and flee when an obstacle is suddenly placed before it. All this takes place at speed because the gradual development of such a capacity conferred a distinct evolutionary advantage on animal life in its locomotive search for sustenance. For the fulfilment of a plant's needs, by contrast, such speed is unnecessary – with the aforementioned exception of the Venus flytrap – and so the development of it would have conferred little to no advantage.

The second key sense in which the animal's movement is qualitatively different from that of the plant is in the control the former has over it. Whereas the differentiated parts of the plant carry out their functions without central direction, the parts of the animal are coordinated by the nervous system. This is most obviously the case when we think of an animal moving from place to place through the sequential movement of multiple limbs, or even those animals that achieve locomotion through a continuous rippling pattern extending through the entire body, such as fish and snakes. Crucially, this central direction does not entail the presence of anything like free will, if we take this to mean a course of action chosen from among various alternatives. Jonas means only that animals – yet no plants – have the ability to act with *intent*. Compare the 'motion with which a cat starts out of sleep in response to some noise' with 'its subsequent actions – alert listening, orientation towards the source of the sound, stealthy creeping in that direction, tense lying in wait for the prey' (*OF* IV: 23). Being an embodied animal ourselves, capable of both types of action, we *perceive* the difference directly: the one kind is closer to vegetative movement insofar as it is involuntary, while the other is intentionally directed towards the world from within.

With this last observation we have arrived at the pivotal moment in Jonas's analysis of animal being. In observing the difference between intentional and involuntary motion we touch upon the connection, first observed by Aristotle, between the animal's locomotive form of life and the presence of sentience and emotion. Jonas likewise suggests that these latter capacities emerge only with the former:

> For it becomes obvious that the 'sentient and appetitive soul' of the animal is not simply a superaddition to a 'nutritive soul' which is the same in plants and animals, but that sentience and appetition belong to a being which is also in its very method of nourishment radically different from the plant; and that the latter is without those faculties because its way of nourishment *allows* it to do without them. (*OF* IV: 49)

Although Jonas argues that locomotion, sentience and emotion came into being contemporaneously, we have tacitly suggested through the preceding discussion that the first of these takes analytic priority. Of the remaining two, sentience is more obviously connected to animal locomotion and so we shall turn to it next.

Jonas points to the inseparability of sentience and locomotion from both angles. From the side of sentience there is the obvious fact that it would be functionally redundant if the organism that evolved to possess it were incapable of acting upon the information it provides (fleeing from imminent danger, for example). Then, viewed from the angle of locomotion, it is apparent that in such action consciousness of two domains is always already present. Firstly the organism must have an 'internal' bodily consciousness: more specifically, an awareness of the 'mutual relations of its parts, the changes produced therein, and the actual process of change itself', which together go by the name of proprioception (*OF* IV: 11). Without such information there is no way in which a locomotive movement can be intended and carried out. Say I wish to cross the road: I must be able to both perceive where my limbs are in relation to the remainder of my bodily form, and sense their successful movement resulting in the carrying forth of my body. Of course, in addition to this internal proprioception, I must also have perception of the external world: that which 'starts where our body is exposed to the action of things other than itself, i.e., on its surface' (*OF* IV: 12). To continue the example just given, I must know by sight and touch where the road is in relation to my body, as well as where the road is in relation to other objects, before I can even conceive of it as something to cross. All this is achieved in the blink of an eye, without deliberation, and yet the action is impossible without it. Thus locomotion is governed by sentience – here meaning both perception and proprioception – through and through.

Now, within the animal kingdom sentience evidently manifests to vastly different degrees: the sentience of the worm is basic compared to that of the starling; the starling's to that of the ape. When contemplating these examples, we are immediately struck by the correlation between degree of sentience and the relative complexity of their various bodily forms. This is indicative of the fact that both perception and proprioception rise to new heights of intensity through the development of specialized body parts, namely limbs, digits and sense organs.

First of all, the development of specific limbs represents a transformation of the organism's proprioception. As a point of comparison, think of the single-celled organism: with its external uniformity, there is no possibility of it perceiving a part of itself in spatial relation to any other (as further away or closer to, for example). Its lack of a central nervous system, moreover, means there is no centre to which its extremities can relate back, allowing them to be perceived *as* extremities. By contrast, the worm represents an advance in both respects: its elongated form entails that it has two clear ends – front and rear – both of which are controlled by a central nervous system, the latter providing a centre that allows the ends of the worm to become ends *for* the worm.

Still, the ends of the worm's body do not constitute limbs, which are jointed appendages of the bodily core. The starling, of course, has exactly these, being blessed with wings and legs. The possession of such freely moving extensions totally transforms the possibilities for proprioception. Distinct from the centre of the body – which can simultaneously move separately, or indeed, not at all – the limb has its particular 'surrounding "medium", which is the situation given for action, and this situation changes as the action proceeds, as part of the success of the action itself' (*OF* IV: 12). Thus the body relates to itself in more complex ways: a limb can be viewed as *here* and moving to *there* – where the latter is projected from the body – but it can also be perceived as 'there' in relation *to* the centre of the body. On the basis of such differentiation the limb can gesticulate and signal, explore open space, make contact with an object and even, in the case of the sparrow's wings, allow for flight.

However, perhaps the decisive moment in the evolution of animal sentience – if we may indeed speak of just one moment – is the development of specific sense organs. This event marks an overturning of the rule by the sense of touch, which, despite taking on a new significance in the centralized animal organism, is nevertheless a continuation of the irritability of a single-celled organism. By contrast, sight – which appears to be the sensory modality to have developed soonest after touch – takes perception of the world to an unfathomably higher level. Although we shall address the full significance that Jonas attributes to sight in the next section, he firstly suggests that sight is 'the indispensable condition for long-range and well-directed motility' (*OF* IV: 27). Although we might note that there are – as always – certain exceptions to this general rule, such as the bat's capacity for sonar, it is indeed true *as* a general rule. In stark contrast to the immediacy of touch, through sight the organism awakens to its environment, perceiving objects that lie at a distance; on this basis it can orient itself towards them and 'strike out into the depth of space' (*OF* IV: 27). As such, the seeing animal is thereby the first kind of organism to fully realize the potential of spatiality that was originally opened up in organismic dealings.

Now, because animals are not only herbivorous but also carnivorous, sense-directed motion changes the conditions of life quite dramatically. The animal can be simultaneously hunter and hunted, predator and prey: thus the 'first great consequence of [animal] motility is the immense sharpening of the competitive element in the interrelations of animals' (*OF* IV: 32). This is particularly true, Jonas suggests, for marine animals, as 'life in oceanic conditions is the most exposed of all' (*OF* IV: 33). He continues:

> The ocean floor being out of reach for the denizens of the surface and the middle regions, there is no place [for marine animals] to hide, no features of landscape with which to blend, no obstruction between self and the eyes of the potential enemies, and therefore no place of retreat even for the periods of rest. [...] Thus it was destined by the aquatic origins of life that the first burst of animal freedom through sentience and motility led into the most pitiless, savage conditions of conflict. (*OF* IV: 33–4)

It should be noted that Jonas does not believe this Hobbesian vision of a war of all against all characterizes animal life in its entirety. But he does argue that it starkly reveals, and indeed shapes, the third key aspect of animal being: emotion, which we shall turn to now.

We noted earlier that well-directed motion and locomotion became possible only as a result of the intensification of sentience through the development of sense organs. Jonas suggests that this evolutionary moment is also 'the birthplace of emotional life', the latter being, he says, a necessary motivation for such action (*OF* IV: 43). Jonas's argument for this claim runs as follows. Sentience, as we have seen, opens the animal up to those beings in the world that either pose a threat to it or contribute to the fulfilment of its needs. This alone, however, is insufficient for the animal to act towards those beings in a way that transcends vegetative irritability. Action also requires an emotional *interest* that serves to drive the animal towards or away from a being. For the animal this interest 'itself is not "perceived", i.e., it is not a function of sentience. Sentience gives the *object* of interest but not the interest itself. The interest is "given" in the form of emotion and passion: it is only in this way that the "objective need" can become an incentive for effort and thereby a cause of action' (*OF* IV: 36; emphasis added).

In terms of the specific forms that such interest takes, spurring the animal to act, Jonas's analysis stresses the hardships that arise with the locomotive form of life. As we have seen, the animal is committed to the active search for nutrition as a consequence of its constitution: thus 'greed' (or, in less moralistic terms, 'want') becomes the first and most fundamental emotion available to the animal (*OF* IV: 57). The second is fear, on the basis of the vulnerability of animal life. Precisely because the animal is both open to the beings that threaten it – predators and natural hazards – and has the capacity to flee those threats, 'dread of annihilation' serves to keep the animal alert and ready for flight (*OF* IV: 35). Lastly, Jonas suggests that the 'third primitive emotion', 'in which the two others are comprised, is that of pugnacity' (*OF* IV: 58). Here the desire to acquire sustenance, territory or a mate collides with the fear of losing out to a competitor, driving the animal into conflict. Far from the Cartesian vision of animals as mere biological machines, lacking an inner life, Jonas concludes that animal 'existence is fitful and anxious [...]. It is essentially precarious and corruptible being, a venture in mortality' (*OF* IV: 59). Here the echoes of Heidegger's 'anxiety' are clear, with Jonas tracing its possibility back to the unique degree of animal freedom.

Although greed, fear and pugnacity are only available to animals (and, of course, humans), Jonas goes on to suggest that the grounds of emotion per se are in fact common to all organisms. Emotion is, he says, 'a more articulated form of the basic self-concern of life as such, a transformation and specification of it according to the level of sentient and motile organization' (*OF* IV: 36). It is, in other words, an advanced manifestation of the care-structure made possible by the two other unique properties of animal being: sentience – which itself has roots in the basic irritability of the organism – and locomotion, which is again a transformation of organismic motility in general. True to Jonas's overarching monism, then, all three phenomena are taken to have emerged from dimensions of vegetative life, even if the qualitative shifts they undergo in the animal raises the latter's being to new heights.

At this point, however, we must raise a fairly significant criticism of Jonas's analysis of animal life, namely that it is individualistic to the point of being solipsistic. We have already observed that his analysis of plant life unduly neglects concern (*Fürsorgen*), and – aside from his references to fear of predation – the same is true of his treatment of animals: sociality features strangely infrequently in Jonas's published works, and is absent altogether from *The Phenomenon of Life*.[2] Leon Kass once raised this very issue with Jonas, who conceded that he had focused on the 'functions of life for an individual animal' as at that time he 'was still too much in the grip of the teachings of Heidegger and his view of life as (mainly) a lonely project over-against-death' (1995: 11). This is an inadequate explanation on Jonas's behalf, however, for two reasons. Firstly, as we have seen, Heidegger identified concern as a component of *Dasein*'s care-structure. And second, and more importantly, it means that Jonas fails to address sexual reproduction, which cannot in any way be considered peripheral to animal being. It is true, of course, that not every animal individual procreates (despite each being the product of procreation), but this is no reason for *The Phenomenon of Life* to ignore the place of sexual reproduction in animal life as such.

Fortunately, however, Jonas *did* broach the issue in *Organism and Freedom*. There he acknowledges, if only fairly briefly, that 'the organic fact we have so far left unconsidered [is] sex' (*OF* IV: 67). Its recognition duly transforms his account of animal life from one of greed, dread and competition to one of coexistence also:

> [W]ith the emergence of more direct co-operation in the sexual act, and again, much later, with the extension of the female role to tending either eggs or offspring, life within the species assumes entirely new features, profoundly affecting the very nature of animal existence. Whatever there is of non-self-seeking traits in the emotional economy of the animal kingdom has its root in this basis of sex and procreation. (*OF* IV: 67–8)

As Jonas says, the world of animals is therefore not simply one oriented towards self-preservation and 'acquisition' but also one which finds fulfilment in 'expenditure' with 'no reference to the metabolic demands of the organism' (*OF* IV: 68).

In addition to sex there is the maternal (and occasionally paternal) animal practice of rearing young, which 'suppl[ies] the foundation of all sociability' (*OF* IV: 69). Here 'sociability' does not mean concern for others, which, as we have seen, may even pre-exist animal life, but rather a 'social life' in the sense of actually living alongside others. This becomes clear when he states that '[w]ith the development of rearing habits intraspecies relations pass from the fleeting nature of the sexual encounter to more durable, either individual or collective, forms of association' (*OF* IV: 69). Rearing allows not only for the animal's parent–child relation to eventually blossom into emotional intimacy – contrast, for example, the life of a penguin with that of a turtle who hatches unaccompanied – but also for the social life of an entire brood, as mentioned, which establishes kinship among contemporaries. From broods the circle of sociability can obviously be expanded to herds, packs, troops and so on, all of which feature structures and norms that eventually bear comparison to human society: the complexity of gorilla and chimpanzee groups being the most obvious example. With reference to natality Jonas is, therefore, able to accommodate the being-with of the animal's world.

Jonas makes one last observation as to the significance of child-rearing: namely that it coincides with the technical skill evidenced in animal life. Although there are exceptions to this – most obviously among primates – Jonas is right to note that rearing procedures account for much of the technical 'measures which go from slight adaptations of existing features in the environment to elaborate artificial constructions' (*OF* IV: 70). Whether an excavated den or a constructed nest, the purpose is usually to account for vulnerability, either of the adult in hibernation or of the child in infancy. This is evidenced by the fact that the majority of such habitations are constructed seasonally – exceptions including, as Jonas notes, the dam-building of beavers and the hives or nests of social insects such as bees and ants, all of which are remarkable in their societal arrangements. He concludes these critical reflections by noting that 'we have in sexuality the root for two extremely important classes of animal behaviour: social and technical', which will eventually take on new forms and significance in the life of human beings (*OF* IV: 71).

Casting an eye back over our reconstruction of Jonas's analysis of animal being, illustrated by references ranging from the humble worm to the great apes, we are struck by a vast existential diversity, far exceeding that which occurs between kinds of plant. This scope of world-openness should not, however, lead us to miss the shared characteristics that mark out animals *as* animals. As we have seen, Jonas follows Aristotle in citing locomotion, sentience and emotion as those very characteristics, which together indicate a single quality that defines animal being: its openness to the world. Plants, whose metabolic and sexual needs are met without recourse to locomotion and directed movement, live in almost immediate commerce with their environment. 'The great secret of animal life', however, 'lies precisely in the gap which it is able to maintain between immediate concern and mediate satisfaction, i.e., in the loss of immediacy corresponding to the gain in scope' (*OF* IV: 45). This mediacy, mandated by the animal constitution, opens up true being-in-the-world for the first time in living nature.

But is our being equivalent *in kind* to that of animals – simply more developed – or is it qualitatively different, since informed by something else entirely? This question has occupied philosophers since the dawn of the discipline, and most, at least prior to the Darwinian revolution, have opted for the latter answer. Identifying what this 'something else' is, however, has proven less than straightforward. And even if we can identify the *differentia specifica* of humanity, can we do so without collapsing into a new kind of gnostic dualism? Let us see how Jonas attempts to solve this riddle.

IV Humans

The final stage of Jonas's *scala naturae* – indeed, the final stage of our reconstruction of his philosophy of life – concerns the nature of humanity and humanity's place in nature. In addressing these issues we move from Jonas's philosophy of biology to his philosophical anthropology, which concludes *Organism and Freedom* and occupies much of the second half of *The Phenomenon of Life*. Although philosophical anthropology has long been a mainstream topic in German-language philosophy, it remains unfamiliar to most Anglophone readers, and so a brief overview of its history, aims and method is required to contextualize Jonas's theories.

As with much German philosophy of the last two centuries, philosophical anthropology has its immediate roots in the thought of Immanuel Kant. Kant's critical philosophy, which stands at the heart of his oeuvre, sought to answer three questions: What can I know? What ought I to do? And what may I hope? Kant suggests, however, that we could reasonably treat 'all of this as anthropology', since each question leads back to a fourth, more foundational one: 'What is man?' (1992: 538). Following Kant, German philosophers of successive generations sought to answer this question by synthesizing the findings of the natural sciences (*Naturwissenschaften*) with the self-understanding provided by the humanities and social sciences (*Geisteswissenschaften*), hoping thereby to obtain the most complete picture of humanity possible.

Two broad approaches developed within philosophical anthropology, representing different perspectives on the same set of key themes. One, which we might call the 'pessimistic' strand, originated in the Romantic-era writings of Johann Gottfried von Herder and found its greatest expression in the early-twentieth-century works of Helmuth Plessner and Arnold Gehlen. All placed emphasis on the ways in which human beings are *deficient* in comparison to other animals. Lacking well-defined instincts, naked, vulnerable and devoid of functionally specialized sense organs and limbs, humanity is, as Gehlen had it, the 'undetermined animal' (1988: 25). To compensate for this incompleteness without, humans were forced to develop *within*: language, rationality, norms and institutions constitute a cultural framework that simultaneously opens up and overlays the physical world in which we move and breathe. Humans are thereby condemned to exist ambiguously, enjoying a superior openness to the world while being, at least in one sense, 'inferior to the animal[,] since the animal does not experience itself as shut off from its physical existence' (Plessner 1970: 37). For this reason we are, according to Plessner, 'the apostate of nature' (2019: 288).

The second perspective – which is by no means wholly incompatible with the first – is, for want of a better term, 'humanist' in orientation. This tradition takes its inspiration from the Aristotelian conception of humanity as the pinnacle of the *scala naturae*. In addition to the sensitive soul belonging to animals, humanity has, according to Aristotle, an additional 'god-like nature' in its capacity to 'think and to be wise' (1984f: 686a). This is its unique character, marking it out from the remainder of life. Identifying humanity's defining feature as a trans-animal faculty of mind – rather than the Judeo-Christian notion of a unique soul – would later become a key motif of the Enlightenment, and find an influential adherent in Kant. According to Kant's anthropology, the *differentia specifica* of humanity is its capacity for reason, both theoretical and practical. In like fashion, the early-twentieth-century philosophical anthropologists Max Scheler and Ernst Cassirer conceived of humanity as having an additional, defining capacity over and above our animality. For Scheler this was our uniquely 'spiritual [*geistiges*] being' (1961: 37); for Cassirer it was our symbolically based culture, earning us the appellation '*animal symbolicum*' (1944: 26).

Despite the great ambition behind philosophical anthropology, and the depth and richness of the work produced by its key figures in interwar Germany, the sub-discipline was eclipsed by the rise of none other than Heidegger. The appearance of *Being and Time* in 1927 struck European philosophy like a bolt of lightning, overshadowing Plessner and Scheler's key works of anthropology which were both published the following year. Then, in April 1929, Heidegger and Cassirer met in the Swiss resort of Davos to give individual lectures on Kant and debate his ongoing relevance.[3] By all accounts, Cassirer's defence of Enlightenment humanism could not withstand the radical questioning of the younger existential phenomenologist. Heidegger was particularly damning regarding the prospects of answering Kant's fourth critical question – 'what is man?' – through philosophical anthropology as it had come to be. The 'deficiency' of philosophical anthropology, as Heidegger saw it, was that it had not been 'expressly grounded in the essence of Philosophy' and yet presented itself as capable of answering the most fundamental philosophical problems: a claim 'whose superficiality and philosophical questionableness jump out at us'

(1997a: 148–9). The existential analytic of *Dasein*, by contrast, addressed 'the question concerning the essence of human beings in a way which is *prior to* all philosophical anthropology and cultural philosophy', and thus constituted the genuinely philosophical enterprise (1997a: 192).

Although he undoubtedly raised important questions regarding the goals and limitations of philosophical anthropology, it must be said that Heidegger's overarching criticism of it was unfair. With his gnostic cast of mind the early Heidegger was blind to the value of establishing humanity's place in nature, and, having construed *Dasein* as merely 'fettered in a body', he had no way of approaching this task even if he had wanted to (1997a: 203). But Jonas, of course, is able to proceed where Heidegger could not. The project of philosophical anthropology, aiming to bridge the natural and human sciences, aligns with Jonas's attempt to interpret biological findings from the perspective of embodied phenomenological reflection. And, just as the former seeks to define humans with respect to other forms of life, Jonas's existential analysis of plants and animals provides him with the equivalent basis on which to carry out such a comparison.

Having dealt with the necessary historical considerations, let us now look in detail at Jonas's anthropological theory. Of the two strands of philosophical anthropology outlined, it should come as no surprise that Jonas is closest to the approach indebted to Aristotle. What is unexpected, perhaps, is the precise form that his anthropology takes. Given that Heidegger had publicly dismissed the sub-discipline in his 1929 debate with Ernst Cassirer, it is ironic that Jonas, who was Heidegger's student until only a year previously, should explicitly endorse Cassirer's version of philosophical anthropology. Atop the *scala naturae*, Jonas claims, humanity is fundamentally differentiated not by reason, self-consciousness or *Geist*, but rather by the element underlying them all: our symbolically grounded existence. Of the various attempts to define humanity Jonas suggests that Cassirer's is correct, stating simply that '*homo = animal symbolicum*' (*OF* V: 37; emphasis added). But on what grounds does he reach this conclusion?

Jonas's investigation is guided by the following heuristic: What is the most basic and widespread physical phenomenon that we would take to be as indisputable evidence of a human creator? Jonas aims to work from an intuited answer to this question back to the faculty responsible for its creation, in the hope that this will lead us to the *differentia specifica* of humanity, and he begins with a thought experiment reminiscent of a work of science fiction:

> The members of an interstellar expedition are being primed for their task of exploring life on another planet. Among their instructions is the particular

one: to ascertain whether there are 'men' over there. They have then of course also to be instructed on how to recognize 'men'. The term 'men' is put in quotes because obviously no likeness of physiological species can be invoked. [...] But it also must be recognizable from without, prior to any communication with the inwardness of the creatures concerned. Thus arises the question of criteria which are neither merely organic nor merely mental. What guides for external discrimination of 'humanity' can the members of the expedition be provided with? (*OF* V: 6)

The extra-terrestrial setting of Jonas's thought experiment is useful in two respects. Firstly, as he notes, it provides a safeguard against judging the presence of the *differentia specifica* of humanity on irrelevant morphological grounds. Second, in searching for an outward manifestation of the *differentia specifica*, we commit ourselves to a phenomenon that is observable, and so open to empirical study.

And yet the setting of Jonas's thought experiment also has its disadvantages. For humanity did not come to be on a distant planet, but here, on earth, in gradual evolutionary distinction from other forms of hominid. Certain morphological features that set human beings apart from our closest relatives will, therefore, be decisive in determining why humanity can act in unique ways. As Jonas himself says, '[T]he increase in man's brain size, his hand, [and] his erect posture reveal their significance in what they allow us to accomplish' (*MM*: 77). Connected to this, the artificial extra-terrestrial setting posited by Jonas necessarily overlooks the continuations between certain human activities and those of our nearest relatives: findings that have an obvious bearing on the question at hand. It is more instructive, therefore, to tie Jonas's quest for a material trace of the *differentia specifica* to the real-world setting of life on earth, where palaeoanthropologists are engaged in that very task. For when we survey the record of pre-history, we find in those sublime depths gradual indications of the human being to come: the creation of tools, the controlled use of fire, the majestic figurative art of Altamira, Chauvet-Pont-d'Arc and – oldest of all – Lubang Jeriji Saléh. To establish which, if any, of these definitively announces the arrival of human beings, we must begin by looking at the criteria Jonas proposes to use in identifying such evidence.

Jonas suggests that any proof of the *differentia specifica* of humanity must fulfil three strict criteria. Firstly, the evidence must be 'unmistakable in itself, i.e., unlike any other phenomenon' (*OF* V: 7). To see why, consider a hypothetical candidate such as apparently decorative handprints and footprints made in the earth. While we might think that this example of aesthetic play reflects something unique about human beings, the resulting prints are *by themselves* ambiguous in this regard: similar prints could have been made completely unintentionally simply by walking or crawling around. Secondly, the proof must indicate an exclusively human trait. This demand rules out both animal instinct and multiple human faculties as possible causes, the latter necessitated by the fact that our search is for the *specific* (i.e. singular) distinguishing feature of humanity. Lastly, Jonas suggests that the evidence must be 'as nearly as possible co-extensive with humanity', and so reasonably expected of human beings regardless of the complexity of their socio-economic arrangements (*OF* V: 7). This criterion is necessary to eliminate any overly demanding or culturally relative candidates for evidence – the building of churches, say – that satisfied the previous two criteria. Together, Jonas claims, these three conditions 'assure that "humanity" is neither attributed where it is not, nor missed where it is' (*OF* V: 7).

Jonas then proceeds to test several candidates for proof of the *differentia specifica* against these criteria: simple tools, the controlled use of fire, tending to the dead, language and image-making. We shall take each in turn and see how they fare.

Firstly there are simple tools: 'A tool is an artificially devised, inert object interpolated as a means between the acting bodily organ (usually the hand) and the extracorporeal object of the action. It is given permanent form for recurring use and can be set aside in readiness for this' (*MM*: 78). An immediate objection to Jonas's definition might be that a tool need not be inert: think of an electric drill, for instance. However, Jonas means any object which is inert either permanently or until it is picked up and put to use, in contrast to a machine which is left to work by itself. Jonas's definition is also intended to preclude any comparison with a tool-like organic extension, such as the web woven by a spider from its spinnerets.

He argues, however, that tools so defined fall foul of the second criterion, as 'here we can most readily speak of fluid boundaries between animal and human capabilities' (*MM*: 79). The oldest and most basic known tools are stones put to the end of hammering, which pre-exist the *Homo* genus itself. We also know that even today this is not a uniquely human accomplishment, as a number of different animal species – principally primates and birds – routinely use stones to hammer. A more sophisticated early tool is the stone handaxe, which, being deliberately fashioned, represents a qualitatively new level of ingenuity on the part of its maker. However, Jonas suggests that palaeoanthropologists treat such axes as indisputable evidence of humanity only because 'we know of no other animal that chips stones', *not* because 'the performance is in itself conceivable only as a human one' (*OF* V: 8–9). This judgement has since proven correct, as we now know that chimpanzees are able to fashion tools from organic matter: stripping a stick for use in drawing termites from a nest, for instance, and even using other tools to make them (Sugiyama 1985: 361). All of this suggests that simple tools cannot serve as the physical evidence of the *differentia specifica* that we seek.

One phenomenon frequently cited as such evidence is the controlled use of fire, which was likely an invention of *Homo erectus*. Archaeologically, the element of control is inferred from the unambiguous remains of hearth-like features, which suggest repeated and so deliberate use. But what does this require on the part of its user? Could a hearth be the creation of an animal as well as of a human, in the same way as the stone tool? Jonas thinks not, holding with the majority opinion that the complexities involved in the controlled use of fire mean that it 'can have originated only in a genuine discovery such as man alone is capable of' (*OF* V: 13). In particular, '[S]uch emotional discipline in dealing with the destructive from which instinct recoils, such judgment of the right measure, and such a span of indirectness between means and end to be bridged by foresight' are beyond the scope of animal being (*OF* V: 13). This is precisely why the controlled use of fire again fails our second criterion (albeit for a different reason than before): it points in multiple directions, drawing on multiple faculties simultaneously, and so cannot function as indicative of the *differentia specifica*.

Perhaps, then, we ought not to look at early instances of technology but other kinds of objective phenomena instead: the entombed or adorned remains of the dead, for instance. Jonas stresses that neither a constructed receptacle for the body nor any grave goods buried with it are to be confused with simple interment of the dead. The latter, as he notes, 'could be an act of mere disposal and thus a habit of social animals' (*OF* V: 14). Indeed, the ethologist Cynthia Moss reports witnessing a group of elephants who, upon encountering the carcass of an elephant not belonging their herd, 'began to kick at the ground around it, digging up the dirt and putting it on the body. A few others broke off branches and palm fronds and brought them back and placed them on the carcass' (2000: 270). If accurately described, this behaviour would indicate that the elephants' being-toward-death extends beyond the merely organismic level, transformed by complex social bonds into a collective tending to the deceased.

Remarkable as this behaviour may be, however, it is nevertheless incomparable to both the creation of a structure to house the dead and the adornment of the body with decorative goods. The presence of the latter 'tells us that a being, subject to mortality, reflects about life and death, defies appearances, and raises his thinking to the realm of the invisible' (*MM*: 85). Sepulchral elaboration, in other words, 'testifies to *beliefs* and embodies *ideas*, and this alone is exclusively "human". [...] Thus funeral customs, where they occur, are telling, and sometimes [the] earliest, evidence of man's transpragmatic spirituality and defiance of sense – something quite different from the adaptive evolution of "intelligence" in meeting the physical demands of the environment' (*OF* V: 14).

Such practices appear to have been the prerogative of *Homo sapiens*, forming part of its complex cultural development known as behavioural modernity.[4] Although this by no means counts against taking burial practices as our material indication of the *differentia specifica* of humanity, Jonas once again suspects that it is too complex a phenomenon. As burial practices embody a custom of ritual and spiritual significance, they presuppose that 'a complex *history* of intra-human relations and of conflicting ideas [have been] reconciled in the authority of a tradition' (*OF* V: 14). This poses a problem, as '[t]hat which is only as a product of history is also subject to the *accident* of history – with regard to the time of its emergence and even with regard to its emerging at all' (*OF* V: 14). It might be the case, in other words, that a given society's mythological and religious ideas about death never took on material form, or else pre-existed that material reflection by considerable lengths of time. As such, although the presence of tombs and burial practices always indicates a human origin, that not all humans create them means that they fail the third criterion set out by Jonas.

The last case has brought us significantly closer to an answer. Unlike tools, ritual death practices are a uniquely human phenomenon, and unlike the controlled use of fire they point towards a single faculty: namely, the possession of abstract ideas. The only problem, as discussed, is that their realization requires too great a degree of cultural development to serve as our sought-for trace of the *differentia specifica*. But there are, of course, less elaborate ways of presenting ideas, the most obvious of which is *language*.

Jonas's high estimation of language is reflected in his statement that the 'philosophy of language must stand in the center of every philosophical anthropology' (*PE*: 265). Language is, first of all, to be distinguished from signification in general. The latter refers to aural, visual or olfactory 'expressions of states of feeling, perception and volition', which are essential to the animal manifestation of concern (*OF* V: 10). We are all familiar with the meaning of the snarl and curled lip of a dog, the arched back of a cat, the chest beating of a gorilla. Even if we had not encountered this behaviour before, it is entirely plausible that we could correctly guess their meanings on the basis of instinct alone. As Jonas says, '*Animal life is expressive*, even eager for expression. It displays itself; it has its sign codes, its language; it communicates itself. Whole rituals of posture and gesture and expressive movement serve the role of signals' (*PE*: 248). Being a kind of animal ourselves, in our interaction with other species 'something passes between us without which there could be no higher understanding, however far it surpasses this elemental substratum' (*PE*: 248).

Of course, communication between humans does surpass this level. In language we have 'a system of sensible and physically reproducible symbols for meanings more or less defined and more or less univocally correlated to them' – in other words, a system of objective signifiers with *abstract* meanings (*OF* V: 10). Whereas the meaning of the dog's snarl lies in its giving immediate expression to an emotion, in language proper the meanings of words refer to ideas that exist independently of a particular invocation. This abstract kind of signification both imbues a great deal of our otherwise animal being and, more importantly, makes the fully trans-animal aspects of human culture possible:

> Rearing of the young means for man essentially teaching them how to speak – by speaking to them. Kinship and authority relations are defined and transmitted through speech. Even our dreams are permeated with words. How much more do words dominate in the life areas indicated by the tool, the image, and the tomb – in planning, work, remembrance, and veneration. And how completely speech-dependent are the worlds of politics and law, and most of all, the relations with the invisible, which nowhere gains form but in words. Man, then, is first and foremost a creature of speech – productive of speech and the product of it. (*PE*: 265)

In language, therefore, we are sure that we have a capacity that both indicates something unique about humans and belongs to each of us, regardless of time and place. Can it then serve as our trace of the *differentia specifica*?

Jonas argues that, despite its clear and overwhelming importance for the being of humanity, language does not fit the remit of our thought experiment. The reason why pertains to the particular forms it takes: written and spoken. The difference between the two extends far beyond the obvious distinction of verbal and visual communication. Speech belongs, Jonas says, 'to the category of doing (πρᾶξις [*praxis*]), writing to that of making (ποίησις [*poiēsis*]); the former is organic action realizing simply itself, i.e., realizing its end in the performance as such; the latter, artificial production (τέχνη [*technē*]) with the result apart from itself – the artefact so produced' (*OF* V: 11). The first problem this raises for our purposes is that only the written word continues to be after the act of producing it has ended, whereas the spoken word ceases to be as the sound of it expires. Had speech and writing developed simultaneously this would not be an issue, as we could simply take the latter as our material indication of the *differentia specifica*. Our second problem, however, is that writing is thought to have developed many millennia after speech, being entirely reliant upon the latter. Requiring, as it does, thousands of years of further cultural development, writing therefore also fails the third of our criteria.

With this last observation we have refined the precise qualities needed in our sought-for phenomenon if it is to satisfy the criteria we have set out. It must be a means of communicating abstract ideas that leaves behind a material trace, and yet not demand the same degree of cultural sophistication as writing. There is only one remaining candidate for such an indication of the *differentia specifica*: namely, pictorial representation.[5]

Jonas defines the image as 'an intentionally produced likeness with the visual appearance of a thing [...] in the static medium of the surface of another thing. It is not meant to repeat the original or pretend that it is the original, but to "re-present" it' (*MM*: 79). For the most part, such images originally took the form of ochre paintings on the walls of caves, and sculptures (such as the famous Venus figurines) carved from stone or ivory. As Jonas says:

> For most readers the cave paintings of Spain and southern France will have leapt to mind. But our evidence does not require the perfection of the Altamira frescoes, or anything even remotely approaching it. In fact, the crudest and most 'childish' drawing, such as dots in a circle representing a human face, would be just as conclusive as the frescoes of Michelangelo. Conclusive for what? For the more-than animal nature of its creator; and for his being a potentially speaking, thinking, inventing, technological, in short, a 'symbolical' being. (*OF* V: 15–16)

Taking the image so understood as our clue to the *differentia specifica*, let us see how it satisfies our three criteria.

Firstly, the production of the image places no unreasonable developmental demands on humanity, either socio-culturally or technologically. Requiring no more than the creation of tools for carving and ochre for painting, the development of such images is a purely cognitive event – one that formed a key part of *Homo sapiens*' behavioural modernity, with the oldest known example of figurative art dating to precisely that period (Aubert et al. 2018). As such, image-making can be justifiably thought of as co-extensive with humanity. Secondly, the image itself is unambiguous. Unlike earlier kinds of symbolic art, which take the form of abstract lines and shapes, the image is instantly recognizable as an intentional attempt at representation. As such, image-making 'from its very beginnings in its most primitive and awkward products, displays a total, rather than gradual, divergence from the animal's [...] – fluid boundaries are not even conceivable here' (*MM*: 79).[6] Lastly, the image speaks to a single, uniquely human property: symbolically grounded existence.[7] In pictorial representation we have the earliest, most comprehensive evidence of humanity's radically new form of openness to the world and itself.

Perhaps the easiest way of approaching an understanding of the image is to note that it serves no immediate biological purpose. As Jonas says, 'No painted venison, however perfect the likeness, will fill a hungry stomach, no likeness of a spear will kill an enemy' (*OF* V: 17). Rather, the chief utility of painting or carving such beings – from the point of the early human – is that doing so grants a purported magical power over the represented thing. Clearly, in the present we do not know how such a magical system is supposed to work, and would not believe in it even if we did. But the idea itself is comprehensible to us nevertheless, and the basis for this understanding is the notion, shared by both the cave painter and ourselves, that the lines painted on the wall depict a real, living being. As such, even if we do not attribute magical powers to a cave painting of a horse, we nevertheless recognize that it is possible for the idealized depiction of the horse to take on such a meaning for its creator. However, this explanation merely invites a further, more foundational question. What permits us to point at the painting and say 'that is a horse', when the painting and the horse itself are such distinct beings – the one made of ochre and the other of flesh and blood?

We have already noted earlier that an image is an intentionally produced representation of something, but this was only a cursory definition. The curious ontology of an image revolves around the term 'representation'. The image of a horse is not – indeed, cannot be – perfectly complete, as if it were then it would not be a representation but rather an actual horse. On the contrary, an image may actually defer from reality in exaggerating or understating certain aspects of the represented being for the sake of emphasis. And yet it can neither neglect nor distort too many core features, as to do so would fail to be an image *of* the thing it purports to represent. Creating an image of something is instead a matter of 'getting hold of the objective features of its appearance' (*OF* V: 19). Jonas suggests that the kind of being that could do so is one with the cognitive ability to apprehend resemblance, or similitude of objective form, at the same time as otherness. This is, he says, 'a statement of essence', and his reasoning is as follows (*OF* V: 25).

The connection between the imaging object – say, lines of ochre on a cave wall – and the being that is represented by it is a third, mediating phenomenon:

the *eidos*. Harking back to the original Greek, Jonas defines the *eidos* as the visible form or appearance of a being. Readers familiar with Plato and Aristotle will recall that the *eidos*, though instantiated in individual beings, pertains to the visual appearance of a kind. Jonas agrees, although sides with Aristotle over Plato in holding the *eidos* to be derived from the perception of individual beings, rather than the other way around: 'Only reality counts, and reality knows of no representation' (*OF* V: 29). Perceiving a horse for the first time we grasp its objective qualities: shape, colour, size, movement and so on. The second time we perceive a horse we recognize it as of a kind: 'this one has a white mane', we might say, where the 'this one' indicates our recognition of the horse as one of two alike beings. Subsequent cases of perceiving individual horses then add to the idea, refining it, all entailing that we concurrently perceive any given horse before us as an individual being while recognizing it as formally like other horses. Thus, as Aristotle argued, our knowledge of the universal arises from the piecemeal stabilizing of that which is given to us in perception, just as an army in flight gradually comes to a halt when one soldier turns and stops, and then another, and another, until all have ceased to flee (1984i: 100a).

Now, Aristotle observed that this is by no means a uniquely human capacity, and Jonas concurs: 'The perception of things as such, and the recognition of the same thing in different instances of sensation, involve [...] a performance of abstraction from the sense "material" with which already animals must be credited' (*OF* V: 31). Only humanity, however, takes the next step of fully abstracting the *eidos* – the shared objective form – from the objects on which it was based. This is precisely what happens in the creation of an *image*. At least in its original manifestations, the imaging object of painting or carving captured the *eidos* of a kind:

> While the recognition of likeness between individual objects still clings to individual existence in each case, the art of the draughtsman detaches the *eidos* from any individual existence, and although he embodies it in an individual existence, that of the picture, this is meant to stand for the *eidos* it exhibits, and through it for all the objects in which it may become individualized. (*OF* V: 49)

Human beings, then, attain a further kind of mediacy in their dealings with beings: whereas animals perceive sameness in beings and on that basis form an *eidos*, humans are also able to mentally detach that *eidos* and give it concrete form in an imaging object. The image of the horse is, in other words, the *visual symbol* for the kind we call 'horse', the image of the hunter the symbol for 'human beings' as they exist in that capacity.

This ability to create physical symbols presupposes – or develops concurrently with – the ability to perceive them as such, that is, as physical bearers of *eidoi* that bear only a superficial resemblance to the beings they depict. Jonas illustrates this with the case of the scarecrow. Both humans and birds perceptually connect the scarecrow to a human being, but for the bird this literally means perceiving the scarecrow *as* a human. Should the scarecrow fail to adequately represent a human being then the bird perceives only rags and sticks swaying in the breeze. Equally, if the bird should gain a better vantage point and see that the scarecrow is not human after all, then the illusion is dispelled, the perceptual connection between it and a human being severed. As such 'either the animal is deceived into mistaking one object for another, or there is for it no relation whatsoever between the two' (*OF* V: 26). Human beings, by contrast, perceive in the scarecrow sameness and otherness at once, recognizing that it is not only a collection of rags and sticks but *also* a symbolic representation of a human being.

With the imaging object pointing away from both itself and the beings it represents, those with the ability to perceive it are granted access to the *eidos* as such as an object of reflection. For 'the early hunter drew not this or that bison but *the* bison – every possible bison was conjured, anticipated and remembered thereby' (*OF* V: 36). Thus 'the bison' as a symbol can take on a host of meanings, inaccessible to the animal that perceives only bison: a symbol of strength, fertility, tranquillity. To be sure, language, which undoubtedly preceded representative art by many millennia, allows us to do the same. The word 'bison' stands for the very same *eidos* drawn from each individual bison encountered. And of course, in speech, song, poetry and mythology the symbolic meanings of 'the bison' are expressed, developed and transmitted through the ages. But the earliest direct evidence of language is permanently lost to time. As such, it is the figurative artworks surviving in European and East Asian caves from around 40,000 years ago that provide us with the earliest evidence of humanity's uniquely symbolic being.

In developing the capacity for abstract thought, humanity attained a *cognitive* freedom of form from matter to parallel the *physical* freedom of form from matter granted by metabolism. Inhabiting a world of symbolic meaning as much as physical sensation, humanity encounters beings in both aspects simultaneously – with the effect that 'in man, objects recede into the farther distance of the generalized and interrelated *eidos* and thus otherness takes on the form of objectivity' (*OF* V: 57). This development signifies what Jonas calls 'the break-through to transcendence, the decisive step beyond confinement to the mere factness of things as they come and demand to be dealt with' (*OF* V:

49). To be sure, for Jonas the organismic mode of being as such is a kind of 'self-transcendence', reaching out from itself into the world of objects (*OF* II: 57). But only humanity transcends not only itself but also the world of objects, attaining a unique way of being that is at once physical *and* symbolic.

Relating to Being as a realm of sense, humanity was then able to undertake the 'symbolical making-over-again of the world', both within and without (*OF* V: 36). Externally, this drew on our ability to make long-term plans at odds with immediate instinct, combined with our capacity to design sophisticated tools. Equally importantly, we were able to hone our movement in accordance with these projected ends: 'In human technology we witness the combined effect of both ways in which the higher perception of man affects his motility. In the range of objects "handled", in the eidetically mediated order of handling them, and in the imposing of envisioned forms on them, we have examples of what motility can do under its new guidance' (*OF* V: 71). Thus the human environment is able to take on an entirely different degree of artificiality to that of the animal. Certain animals, such as beavers, are instinctively guided to construct sophisticated habitations that far exceed in degree of artificiality the cave occupied by early humans. But only the latter being, with its capacity for abstract and long-range eidetically grounded thought, had the potential to one day undertake construction of the immense cairns, dolmen and stone circles that survive from the Neolithic era, let alone to create the global technological society of today.

Only human beings, moreover, would have *reason* to carry out such biologically unnecessary tasks, and this is equally due to humanity's openness to the *eidos* enriching its inner life. We have just indicated the decisive role of eidetic imagination in invention and controlled movement, but we can also attribute our highest faculties to its development. Since the *eidos* 'has to be more or less true to the object', the former can be compared against the latter (*PL*: 172). This gives rise to a rudimentary experience of truth and falsehood as correspondence, since sensation can either accord with or contradict eidetic beliefs possessed by humanity.[8] Through the *eidos* we thereby have at our disposal 'the enduring image of things, which [humanity] henceforth owns as *knowledge* of the object in question' (*OF* V: 41). The reverse act of comparison, with the *eidos* serving as a benchmark, is involved in the ability to disinterestedly observe change over time. From this Jonas draws a striking conclusion: the *eidos* lays the ground 'for some basic concept of philosophy' by making explicit 'the contrast between change and the unchanging, between time and eternity' (*PL*: 152). Thus our symbolic existence opens up the inner world, the life of the mind, at the same time as it allows us to project what was inner on to the outer world in the form of eidetically guided action. In short, 'There cannot be rational man, technological man, aesthetic man, religious man, metaphysical man, political man, scientific man, without pictorial man' (*OF* V: 37). Thus the original *Homo sapiens* who left images behind are the first of whom we can definitively say: here true humanity existed.

V Being is one

Having reconstructed both parts of Jonas's philosophy of life – the existential analytic of the organism and the *scala naturae* – we are now in a position to evaluate its success. Recall his overarching thesis: '[E]xistentialism, obsessed with man alone, is in the habit of claiming as his unique privilege and predicament much of what is rooted in organic existence as such.' Recall also his method: 'an "existential" interpretation of biological facts', the latter ranging from metabolism through to the palaeoanthropological record. Jonas's philosophy of life proves itself, therefore, to the extent that its *terminus a quo* unites with its *terminus ad quem*. That is to say: because Jonas's starting point is existential-phenomenological reflection, and because he seeks on that basis to undertake a reading of the biological realm from its simplest beginnings through to humanity, it ought to culminate where it began: human existence. Let us see, then, whether Jonas is indeed able to trace the 'privilege and predicament' of being human, described by Heidegger in his existential analytic of *Dasein*, back to the features of the organic realm.

We suggested in Chapter 1 that Jonas holds thrownness, fallenness and authenticity to be contingent on culture, while in Chapter 2 we saw that Jonas traces much of the remainder of the care-structure back to the being of the organism as such. And, in the present chapter, we have observed that a rudimentary manifestation of concern could be discerned in the behaviour of plants and animals. This still leaves, however, the important question of the existence of *Dasein*: that *Dasein*'s being is an issue for it, and, having to take a stance on its own being, typically does so according to the culture it inherits.

How, then, does Jonas account for existence: man's being 'concerned with what he is, how he lives, [and] what he makes out of himself' (*PL*: 186)? Jonas's answer again lies in the symbol. As we have seen, humanity is opened up to reflection on and knowledge of beings through the ability to capture the *eidos* in word and image. The majority of early human art depicts mammalian life encountered beyond the walls of the cave – the horse, the bison, the lion and so on – and quite understandably so, since these social animals are so close to us in important respects and yet different in others. But we also find depictions of human beings themselves, and not only in painting but also in the remarkable 'Venus figurines', the oldest of which – the Venus of Hohle Fels – similarly dates to the period of behavioural modernity. Here we have the earliest indication of humanity reflecting upon *its own eidos*, that is, evidence of the possession of true self-consciousness. Here a 'new dimension of *reflection* unfolds, where the *subject* of all objectification appears *as such* to itself and becomes objectified for a new and ever more self-mediating kind of relation' (*PL*: 185). With this development the nature of humanity becomes a problem to itself: '*Quaestio mihi factus sum*: "a question I have become unto me"' (*PL*: 187). And Jonas suggests that in the tomb, adorned with symbolic artefacts, this existential questioning 'takes on concrete form: "Where do I come from; where am I going?" and ultimately, "What am I – beyond what I do and experience at a given time?"' (*MM*: 83).

Hence we arrive at an understanding of ourselves in being that we have to take a stance on, as Heidegger so vividly described. As Jonas says, 'Man models, experiences, and judges his own inner state and outward conduct after the image of what is man's. Willingly or not he lives the idea of man – in agreement or conflict, in acceptance or defiance, in compliance or in repudiation, with good or bad conscience' (*PL*: 185–6). Living in the light of our own *eidos*, we are capable of abiding by the virtues we hold and flourishing according to the account of the good we subscribe to. We are also, however, capable of constructing *false eidoi* of ourselves, and into this category Jonas places existentialism's 'worldlessness' in general and Heidegger's throwness and fallenness in particular (*WPE*: 31). We shall see in Chapter 5 that Jonas holds that humanity may nevertheless divine its *true* image. But whether it happens to be true or false, '[t]he image of man never leaves him, however much he may wish at times to revert to the bliss of animality' (*PL*: 186).

With the last fundamental component of Heidegger's existential analytic accounted for, we conclude our survey of Jonas's philosophy of life. His existential interpretation of biological facts involves, as we have seen, not only a rejection of the Cartesian bifurcation of nature but also the latter's contemporary progeny – including, ironically, a sizeable part of Heidegger's existentialism. All this amounts to a thoroughgoing refutation of the gnostic tenet that the mind, or soul, is distinct from matter. Weaving the threads of matter and mind back together, Jonas shows that, on the contrary, humanity represents the pinnacle of a tendency in Being towards freedom and world-openness. Put epigrammatically: where gnosticism holds that Being is two, Jonas argues that Being is one.

His ontology is not only theoretically valuable, however, but also practically useful, as the overcoming of gnostic metaphysics allows us to restore to nature its proper share of dignity and value, and on this basis act accordingly. But what sorts of value belong to human and non-human life respectively? And what kind of ethic does this yield – one that remains anthropocentric, or one that casts off this legacy in favour of a biocentric or even ecocentric orientation? These are our topics for Chapters 4 and 5.

4

Values and the good

I The axiological dimension of teleology

The present chapter is concerned with the issue of axiology in Jonas's thought. But what is axiology? The term, which is not widely used in Anglophone philosophy, refers to the study of value (from the Greek, *axia*), and is thereby distinct from ethics or moral philosophy, which concerns character and principles of conduct: the latter being our topic for Chapter 5. Here we are interested in what sorts of values there are in the world, which beings possess them and whether, drawing on Jonas's references to species and the *nisus* of Being, it makes sense to speak of values held not just by individuals but also by collectives, and even by the biosphere itself.

As with Jonas's philosophy of life, his axiology is formulated in response to modern gnosticism. We saw in Chapter 1 that the second central tenet of modern gnosticism is the conviction that nature itself possesses no value, positively or negatively: it simply *is*, and any value to be found in the world is instead a subjective human evaluation of it. From this belief follows the purported freedom to treat non-human life as we please: after all – so the argument goes – if nature possesses no good of its own, then it cannot be wronged, and so makes no ethical claims of us. But this argument implicitly rests on the materialist conception of life as devoid of ends, which we have duly discarded. Having done so, the search for value in non-human nature, and the demonstration that life *does* matter, may be undertaken. We shall see that although Jonas is unable to combat gnosticism's nihilism to the desired extent, we are nevertheless able to reveal a far richer and greater ream of value in Being than modern thought typically admits of. This is enough, we shall suggest, to underpin his ethic of responsibility introduced in Chapter 5.

Compared to Jonas's ethics and philosophy of life, however, his theory of values is relatively under-developed: important distinctions and steps in the argument are not always made explicit, or are given too little attention. For this reason we shall attempt to interpret Jonas's axiological claims by situating them within the relevant ethical debates, in particular – following Theresa Morris's lead (2013: 96–109) – those of environmental ethics. The reason for the latter choice is not only that Jonas was a pioneer of environmental philosophy but also because in that sub-discipline the issues we are here concerned with have been pursued most radically. Take, for example, the question of which entities have the status of ends-in-themselves, the answer to which is commonly thought to inform – or even determine – the proper scope of direct moral consideration. Western thought has historically tended to attribute this distinction only to human beings (or even just a subsection thereof). In recent decades, however, it has become more common to grant at least sentient animals this status, and thus direct moral considerability. This already represents a departure from our anthropocentric past, but environmental ethicists have pushed further in an attempt to account for the value of living beings as such: an axiological 'biocentrism'. Yet other environmental ethicists take the next step, and argue that entire ecosystems are ends-in-themselves ('ecocentrism'), or even that the earth as a whole counts as such ('holism').

In the secondary literature there is little consensus as to where Jonas's thought is situated on this widening axiological spectrum. Some commentators believe that he advocates 'following nature' (Krebs 1999: 99), indicating an ecocentric or holist position, while others claim that his thought has 'anthropocentric tendencies' (Attfield 1991: 202). In an attempt to account for this dichotomy, Lawrence Vogel argues that Jonas manages to 'have it both ways' by 'undercutting the very distinction between anthropocentrism and non-anthropocentrism' (1995: 37). Evidently some conceptual clarification is required to establish which of these interpretations is correct, and why they can all seemingly claim textual support.

Our investigation begins from the findings of Chapter 2. There we identified immanent teleology, or goal-directedness, as definitive of the organism, and this in two forms: self-organization and behaviour. To begin with the latter, recall that Jonas characterized each organism's being-in-the-world as care. The specificities of care are in each case determined by the needs of the organism, but we noted that, at least initially, the organism's engagement with beings takes the form of a means–end relationship: entities are encountered as something-for-nutrition, something-for-traversing, something-for-shelter and so on.

The structure of this relationship is critical. First, note that as the end is served by the means the latter is subordinate to the former. Second, note that being a means is not inherent to the thing itself, but rather a status applied by the entity whose ends are to be satisfied through it. To be sure, there must be something about the thing such that it affords the possibility of being taken up as a means for a particular end, but the actualization of this means-potential is determined only by an entity that has such ends. Taken together, these factors entail that a being assigned the role of a means acquires its significance *qua* means to the extent that it acts – or fails to act – in satisfaction of the end to which it is put. Imagine, for example, a butterfly seeking shelter from the rain. If it should take cover beneath foliage riddled with holes from a leaf-feeding caterpillar, such that rainwater dripped onto the butterfly, then the leaf in question would fail in the role assigned to it as a means. What can we learn about value from this? The axiologically crucial word here is 'fail', for it represents an outcome *for* the butterfly courtesy of its goal-oriented encounter with the leaf. Of course, this failure emerges only in and with the butterfly's encounter with the leaf: from the point of view of a neutral observer the leaf itself remains just as it is, neither more nor less.

Now, it might be supposed that this goal-derived outcome is dependent upon the consciousness of the butterfly, such that absent the latter there would be no possibility of the former. Presumably it is indeed the case that the butterfly, as a sentient being, will perceive the outcome of its encounter. But this represents not the creation but rather the *mediation* of organismic goals and their outcomes, the latter pre-existing the perceptual faculties. Even a living being that lacks the capacities we attributed to animals in Chapter 3 would still, *qua* organism, possess ends that are no less real for being 'silhouetted in premental form', as Jonas says (*IR*: 75). It is on this basis that beings encountered by the non-conscious organism in acts of care either succeed or fail *for* the organism. To pick a striking example from nature: the 'zombie fungus' *Ophiocordyceps unilateralis* kills ants of a particular genus by infecting the body and manipulating the host's behaviour. The ant is driven to climb the stem of a plant and, using its mandibles, secure itself to the vein of a leaf. The host thereafter dies as the fungus grows spores from inside the body of the ant outwards, through the head first. Clearly, this grisly death is a terrible fate for the ant. But for the fungus, the ant successfully serves its purpose as a means to the end of nutrition, even though there is no doubt that the fungus lacks the sentience necessary to perceive or feel this fulfilment.

All organisms, in short, encounter other entities as means, allowing for an outcome that depends on the degree to which their ends are met. For conscious organisms this will presumably be experienced as an evaluation, mediated by sentience and emotion; the non-conscious organism, lacking these modes, will be largely blind to the fulfilment or failure of its ends. But the outcome is there for it nevertheless, and the axiological consequence is this: something can be *better or worse for the organism* according to the satisfaction or otherwise of its ends. As Jonas says, '[W]ith any *de facto* pursued end […], attainment of it becomes a good, and frustration of it, an evil; and with this distinction the attributability of value begins' (*IR*: 79). As Jonas's statement indicates, this 'better or worse for' entails, as a matter of logical necessity, that the organism has a good of its own. It is good for the fungus to infect the ant, and bad for the ant to be infected. Regardless of whether or not the organisms in question feel the fulfilment or frustration of their ends, the mere fact of having ends means that they have a good. To be clear, nothing yet said about an organism's good implies judgement as to its objective goodness or badness, which is a separate issue that we shall turn to later. Rather, the good in question is strictly subjective, '*for* something' (*IR*: 52).

At this point one might ask whether machines – which, as we have seen, are also teleological beings – equally have a good of their own. For example, in Chapter 2 we discussed the teleological functioning of a car: Can we not say, in light of the analysis just given, that it is bad for the car to break down? Jonas's response is that malfunctioning could very well be bad *for its user*, but not for the car itself. We said in Chapter 2 that to run properly the car requires petrol, oil, water and so on, yet this requirement does not represent a need on the part of the car, which alone would entail a good of its own. The reason is as follows. The car's material identity is complete whether switched on or off, because the formal division between itself and the world is not its own doing. Thus whatever it requires to run properly is not based on any kind of care, grounded in a processual existence, but merely represents an input into a fixed and self-sufficient system. And the reason for this self-sufficiency harks back to the nature of the machine, which is an arrangement of matter in accordance with a *telos* that transcends it. The upshot is that only the designer who instilled that *telos*, or the user who appropriates it, is served by the functioning of the car.

Behaviour was, we recall, only one type of immanent teleology belonging to the organism. Turning now to the second, self-organization, we shall see that it allows us to deepen the notion that each individual organism has a good of its own. As explained in Chapter 2, the organism formally sustains itself in the act of metabolism. Through the latter process parts of the organism are reconstituted, from the simple such as ectoplasm and the cellular membrane, through to the complex like muscle, fat and organs. All together form a systematic whole of interdependent functions and processes: as Jonas says, 'Every organ in an organism serves a purpose and fulfils it by functioning [...] everything is *de facto* so arranged that in effect it contributes to the maintenance and performance of the whole' (*IR*: 65). Thus we concluded that the whole is both more than the sum of its parts and generative of them as parts.

Once again, this immanent *telos* entails a value. Since the parts of an organism are subservient to the life of the whole – constituting, in other words, means to the end of continued existence – we have a value-criterion by which to judge their performance. Take the heart, for example.[1] Its function is to pump blood around the body. But its purpose is that this function aids the organism in its pursuit of continued existence, which then allows for an outcome: the heart either successfully fulfilling its role as a means to the end of the life of the organism, or failing to do so. Note that, once again, the good of the organism here exists prior to any psychological valuation: having a working heart is good for the organism whether that individual actually wishes to be healthy or not. For all living beings, health – and its opposite, illness – pertain to a goal-derived outcome that pre-exists conscious positing. And, as before, the step from value to subjective good is clear: the organism has a good of its own not only in fulfilling its goals that reach out into its world, but also in fulfilling the goals immanent in its very constitution.

The move from teleology to subjective good represents an advance in Jonas's attempt to overcome modern nihilism. We can refine his theory, however, by differentiating between the different *sorts* of values that follow from natural teleology. An appropriate distinction to introduce at this juncture is that of instrumental and intrinsic value, which has preoccupied environmental ethicists since the development of the sub-discipline.

While no consensus exists in the literature as to the precise meaning of the terms, Jonas's usage of instrumental and intrinsic value aligns with their most common definitions. *Instrumental* value refers to something's use-value as a means to an end posited by the valuer: whether my coffee cup is comfortable to grasp and holds the liquid without leaking, for example. By contrast, *intrinsic* value is that which we find in something disinterestedly, without reference to instrumental needs. When we contemplate the beauty of a natural landscape, or the cherished company of a loved one, we realize that these are not valued as means to an independent end, but rather for what they are in themselves. They just *are* valuable to us. Note, however, that intrinsic value, like instrumental value, does not exist in the valued being independently of the valuer. Someone who could only value the magisterial giant sequoia in instrumental terms might strike us as a philistine, yet we could not say, alas, that such a person was objectively ignorant.

Which beings, then, are according to Jonas's philosophy able to value other beings either instrumentally or intrinsically? The examples of instrumental and intrinsic valuing just given are drawn from human experience, but such valuing may just as plausibly belong to all living beings. Firstly, instrumental valuing appears to belong, as an axiological correlate, to both behaviour and the part–whole relation inherent to self-organization, as in each case something serves as a means to the ends of the organism. The animal that hunts and seeks out a mate, and the plant that draws energy from the sun and soil, both instrumentally value the beings that serve their ends. Likewise, a working heart and a sturdy stem serve the whole animal and whole plant, respectively. We may go further, however, and note that Jonas's account of behaviour and self-organization also entails the presence of intrinsic valuing. Why? The reason, as Jonas says, is that self-organization – and frequently care also – ultimately serves to sustain the life of the organism and its particular mode of being. As he puts it: 'To secure survival is indeed one end of organic endowment, but when we ask "Survival of what?" we must often count the endowment itself among the intrinsic goods it helps to preserve' (*MM*: 93). That is to say: because the organism seeks survival for its own sake and as the particular being that it is, its existence is thereby valued 'beyond all instrumentality' (*MM*: 93).

With this latter move we arrive at a pivotal moment in Jonas's axiology. As the organism is organized to ensure and frequently acts in the service of its continued existence – which means not only its bare existence but also the *way* in which it exists – he concludes that the organism is an 'end in itself' (*IR*: 56). Each life form enjoys the status of end-in-itself not because of its capacity to feel pleasure and pain, as utilitarians would have it, or its ability to uphold the moral law (as in Kant), but rather because its immanent-teleological constitution quite literally entails it. The organism is, uniquely among beings, '*its own end*' (*IR*: 56) Thus far, therefore, we may say that Jonas's axiological position is as follows: instrumental and intrinsic valuing are the axiological dimensions of teleology, and these in turn entail that the organism enjoys the status of an end-in-itself. Evidently this takes us far beyond the confines of anthropocentrism and into a biocentric axiology.

II Species and the biosphere

We have now shown how Jonas's philosophy of life bridges biology and axiology via teleology at the level of the individual organism. However, some environmental ethicists have argued that species, ecosystems and even the biosphere as a whole have values that transcend the merely instrumental. Although Jonas does not make sustained arguments for any of these positions, there are enough hints scattered throughout his work to suggest he was at least sympathetic to the notion that species and the biosphere could have a greater axiological significance than that of mere instrumental value for us. In order to establish what he might have meant, we shall again locate his comments in the debates in environmental ethics over the possible attribution of value to these collectives.

References to species are infrequent in Jonas's published works, which typically focus on the individual organism or life as such. Yet his claim that 'wanton and needless extinction of species' is 'a crime in itself' suggests that the idea is, for him, imbued with normative significance (*TSE*: 894).[2] Why this is the case is not fully explained by the context. However, in an unpublished manuscript Jonas offers the following observation, which introduces a different sort of value than that discussed so far:

> Now, something can be termed 'good' by its own intrinsic standards, unrelated to anything else and regardless of my likes or dislikes: for instance[,] this living body – snake, bug or bear – if complete in its proper parts, all in good working shape, each doing its proper work in proportion to the others and the whole [...] is then a 'good' specimen of its species, of which there can also be impaired, imbalanced or disordered specimens. I may wish the whole species extinct and must still grant that by its *internal criteria of wholeness and excellence*, this happens to be a very good representative of it. (*WGM*: 2; emphasis added)

At first the quotation appears to reiterate the idea, mentioned in the previous section, that the individual organism is an end-in-itself by virtue of its teleological structure. However, Jonas then moves to the notion that the structure of the individual is in accordance with that of the species, and on this basis can be judged as a good or bad of its kind. We shall attempt to make sense of this.

The idea that each species exists in a way that is particular to it, providing us with a standard by which to judge the individual specimen as a good (or bad) of its kind, is, essentially, an Aristotelian notion. Aristotle tells us in the *Nicomachean Ethics* that '[t]he excellence of a thing is relative to its proper function' (1984d: 1139a). For Aristotle this applies to us as both organismic and social beings: if I am someone's friend, there are criteria by which to judge whether I am a *good* friend or not; if I am a chef, there are criteria by which to judge whether I am a *good* chef or not and so on.

According to the environmental ethicist Paul Taylor, the normative concept at work here is merit, which Taylor describes as 'apply[ing] grading or ranking standards' to the individual and thereby determining 'whether it has the "good-making" properties (merits) in virtue of which it fulfills the standards being applied' ((1986) 2011: 130). Jonas himself makes room for this in the quotation with his reference to 'a "good" specimen of its species, of which there can also be impaired, imbalanced or disordered specimens'. Again, as with the goal-directed behaviour of an individual organism, the invocation of a good of a kind makes no claim to objective goodness. For instance, a good criminal is one that commits a crime and evades capture, though we are loath to say that being a criminal is itself good. In the case of an organism Jonas likewise suggests that, regardless of any objective goodness pertaining to the existence of the species, if it exists then a species-valuation is always present.

The problem with this argument is not normative but ontological, pertaining to what a species actually is. While few people – unless they doubted the existence of the external world altogether – would maintain that organisms do not exist independently of our conceptualizations of them, with species it is not so simple. Are we looking for a class or category that is really *there*, or just one more or less contingently imposed by us on the phenomena? Aristotle had argued that species were not only natural but also unchanging kinds, and it seems that with Jonas's reference to an 'internal criteria' of wholeness and excellence he here follows suit. But as indicated at the beginning of Chapter 3, after Darwin this is no longer an option. As Jonas himself puts it elsewhere, species are not fixed but merely 'relatively stable, and [...] this stability represents only the temporary equilibrium among the forces which generally determine the structure as successful' (*PL*: 50). Having acknowledged this it becomes difficult to see how Jonas could argue for an internal criterion of species-valuation, since the standard by which we are to judge the merit of individual specimens is lacking.

It is possible, perhaps, to more charitably interpret Jonas's claim by appealing to a non-Aristotelian understanding of species as natural kinds. Accommodating evolution by natural selection, such a theory would introduce a temporal or historical dimension to the definition. This is the inspiration for the biological (as opposed to the Aristotelian 'morphological') species concept first advanced by Moritz Wagner, and later by Ernst Mayr. According to Mayr's version of this definition, 'Species are groups of interbreeding natural populations that are reproductively isolated from other such groups [...] every species is the product of evolution, or more precisely the product of speciation, and [...] must have certain qualities that are the consequence of that history' (1988: 318). While it might be objected that Mayr's reference to 'natural' populations unduly complicates matters, for our purposes the introduction of a historical criterion allows for the concept of species to take on a temporal dimension. On this definition we could say that a group of organisms does constitute a natural kind, the boundaries of which are demarcated by reproductive isolation from their evolutionary relatives.

The problem with this account of species, at least for Jonas's purposes, is that the significant phenotypic differences it permits undermine the idea of an internal criterion of goodness. One wonders how such a criterion could belong and apply in equal measure to both polar and grizzly bears, for example, or lions and tigers. Even though these pairs are capable of interbreeding and producing fertile offspring, thus counting as single species according to the definition, the form and behaviour of the polar bear are distinct enough from that of the grizzly, and the lion's from that of the tiger, to render the notion of a shared species-specific good implausible. A post-Darwinian conception of species such as Mayr's cannot, in short, uphold the Aristotelian notion of a species essence that alone could provide the basis for a natural good of a kind. This ontological quandary is probably why the notion of species-valuation remains undeveloped in Jonas's writings, at least compared to the 'hard order of ecology', which we turn to now (*IR*: 137).

As stated, Jonas's greater concern is with the biosphere, references to which are scattered throughout his work from the late 1960s onwards, but which features particularly prominently in *The Imperative of Responsibility* (*IR*: 6–8, 136–40). The biosphere has a conceptual advantage over species as it is rather more easily identifiable: it is the totality of all living things on planet earth and the relations between them. This is not to say, of course, that there are no vital connections between the biosphere and non-living entities such as the atmosphere, or extra-terrestrial beings like the sun, upon both of which the biosphere's existence is dependent. But the boundaries of the biosphere itself are clear, and – helpfully for our purposes – as a collective it is axiologically comprehensive, referring to all known beings that have the status of ends-in-themselves.

For the Greeks the structure of the world and the wider cosmos was thought comparable to the structure of an organism: as Jonas explains, 'To the same pattern corresponds every entity in nature, if in lesser degrees of completeness and self-sufficiency. Each is part of a greater whole, an end in itself and a whole for its parts' (*PL*: 95). However, he insists that with the benefit of greater knowledge of ecology and natural history we must refute this '"Aristotelian" idea of a safe teleology of "Nature"' (*IR*: 138). In other words, just as we are forced to abandon the notion of the ongoing essence of a species as anachronistic, so too are we led to the conclusion that there is no permanence or intrinsic harmony to the biosphere. The reality is rather that

> encroaching on other life is *eo ipso* given with belonging to the kingdom of life, as each kind lives on others or codetermines their environment, and therefore bare, natural self-preservation of each means perpetual interference with the rest of life's balance. […] The sum total of these mutually limiting interferences, always involving destruction in the particulars, is on the whole symbiotic but not static, with those comings, goings, and stayings known to us from the dynamics of prehuman evolution. (*IR*: 137)

Most radically of all, the entire biosphere's symbiosis is disrupted with the advent of modern technological civilization, itself, through us, a part of the biosphere. As Jonas memorably puts it, 'Nature could not have incurred a greater hazard than to produce man' (*IR*: 138).

Where, then, does non-instrumental value fit in this picture of an antagonistic and self-imperiling biosphere? Jonas points towards an answer in that the interaction of living beings comprising the biosphere, though often hostile for the individual, is nevertheless the framework that allows such beings to exist. This constitutes one of life's basic tensions: some individual organisms' ends are made subordinate to the 'more comprehensive ends of the biosystem' so that life as a whole might flourish (*IR*: 235). Taking Jonas's invocation of ends here as literal, we can flesh out his axiology by comparing it to that of Holmes Rolston III. In this instance the relevant concept is that of 'systemic value', which Rolston defines as a 'productive process; its products are intrinsic values woven into instrumental relationships' (1988: 188). By producing valuing beings and their interrelationships – however antagonistic these may be – the biosphere appears to perpetuate its own existence, presupposing that it is an end-in-itself.

This raises a moral worry, however, regarding the subordination of the individual to the collective. For example, Rolston says that '[t]he objective, systemic process is an overriding value, not because it is indifferent to individuals but because the process is both prior to and productive of individuality' (1988: 191). Here he appears to pre-empt and counter the charge of eco-fascism, or at the very least 'a detached indifference to individual welfare' (Callicott 1984: 303). This is a genuine concern for the simple reason that attributing systemic value to the biosphere, which requires death and a fortiori suffering to function, entails that organisms otherwise deemed to be ends-in-themselves are made subordinate to the greater good. And unfortunately Jonas does endorse such a position. Rejecting a 'sentimental' approach to the interdependencies of the biosphere, he claims that 'to eat and be eaten is the principle of existence' (*IR*: 137). Even if we accept the Heraclitean principle that '*polemos* – war – is the father of all things', as Jonas does in this context, the identification of its codification in the biosphere as an end-in-itself makes us decidedly uneasy (*OS*: 23).

This concern aside, perhaps the most pressing problem with the notion of systemic value is whether or not it is actually a meaningful concept. Apparently without realizing it, Jonas's work on the organism undermines the idea. The reason is that, in contrast to even the lowly amoeba, the biosphere is not itself a purposeful being but only a *collective* of such beings. Put another way: unlike an organism the biosphere is not self-organizing, and thus generative of its parts, but is rather reducible *to* its parts. It is true that some ecologists have compared the biosphere to an organism – most famously James Lovelock, whose Gaia hypothesis conceives of the earth as a 'super-organism' of individual organisms in much the same way as the Greeks. As indicated, however, the likeness between an organism and the biosphere is merely superficial. Although an organism consists of parts, their *arrangement* is the work of the organism: as Jonas says, '[T]he membership of elements in an organism is an achievement of the latter for its sake' (*OF* II: 33). This accounts for the high degree of formal cohesion among parts, standing in stark contrast to the largely autonomous organisms comprising the biosphere, which are not ordered by the earth, and do not act for *its* good but rather their own.

This raises the question of whether any good of its own can be attributed to the biosphere that is not simply an aggregate of those of its component parts. Rolston certainly thinks it can, claiming that '[t]he value in this system is not just the sum of the part values' (1988: 188). But given that the biosphere is only a collective it is hard to see how this can be. The good of a family, or a nation, or any other social group is a common good formed of an aggregate of its individual components' goods. We cannot say that there is some sort of greater, hypostasized good belonging *to the group as an end-in-itself*, and the same is true of the biosphere. The latter is instead only instrumentally good for the beings that compose it, as Robin Attfield remarks:

> Certainly everything which is of value (and located anywhere near our planet) is located in the biosphere, and the systems of the biosphere are necessary for the preservation of all these creatures. But that does not give the biosphere or its systems intrinsic value. Rather it shows them to have instrumental value, since what is of value in its own right is causally dependent on them. (1991: 159)

This seems unanswerable. Accordingly, an account of the systemic good of the biosphere, based on what Jonas referred to as its 'ends', must be rejected.

This cursory look at two further types of good hinted at in Jonas's work, and elaborated upon via the comparable theories of various environmental ethicists, has led to something of a dead end. We have seen, firstly, that attributing merit to an individual with reference to the collective good of its kind makes logical sense, but seems to lack a sufficient ontological foundation. Conversely, the reality of the biosphere is uncontroversial: there are indeed living beings on planet earth that impact upon one another, the sum total of which can be called the 'biosphere'. This, however, is a mere collective rather than an immanently teleological system, and so cannot be conceived of as an end-in-itself.

III The good of Being

Through the positive arguments of the first section and the negative conclusions of the second, we have arrived at a concrete position: namely, a biocentric axiological theory. This is enough to refute the existentialist doctrine that humanity is alone in an uncaring world with its projected meanings. Rather, humans are situated atop the scale of living beings which all share in the condition of valuing their continued existence, and possessing the status of ends-in-themselves. As Jonas succinctly puts it, 'Nature harbors values because it harbors ends and is thus anything but value-free' (*IR*: 78). However, the demonstration of this does not represent a *victory* over modern nihilism – as Jonas notes, 'no obligation can be derived' from the discovery of values in nature, which 'seem to enjoy no other dignity than that of mere facts' (*IR*: 79). Any ethic appropriate to technological civilization built on this basis cannot, therefore, be regarded as objectively binding.

To overcome this limitation Jonas attempts to demonstrate the objectivity of his axiological findings. Unfortunately his attempt to do so – though admirably bold – is the weakest aspect of his philosophical system. Jonas freely confesses to running 'head-on against the stone wall of two of the most firmly entrenched dogmas of our time: that there is no metaphysical truth, and that no "ought" can be derived from "being"' (*OG*: 51). While commendable, this has predictably exposed him to criticism, in particular from his German interlocutors. Karl-Otto Apel, for example, claims that Jonas 'reaches back behind Kant' in his argumentation, reverting 'to a religio-metaphysical belief that is incapable of a rational foundation' (1996: 225, 219). Similarly, Ullrich Melle claims that Jonas 'falls back on a pre-transcendental, objectivist metaphysics', although Melle concedes that this recourse 'is not completely uncritical, i.e., dogmatic [...]

since Jonas does not claim ultimate justification or absolute truth for it' (1998: 340–1). Perhaps most straightforwardly, Wolfgang Kuhlmann laments Jonas's 'unsuccessful philosophical foundation of the proposed basic norms' (1994: 282). We shall see that although there is some truth in these accusations, Jonas's movement from subjective goods to objective good is not to be dismissed out of hand.

Of the two philosophical faux pas to which Jonas confesses the more problematic is his rejection of the is–ought gap: Hume's observation that a normative conclusion cannot be logically derived from premises regarding the way the world is. The famous passage from *A Treatise on Human Nature* is worth quoting in full:

> In every system of morality, which I have hitherto met with, [...] I am surpriz'd to find, that instead of the usual copulations of propositions, *is*, and *is not*, I meet with no proposition that is not connected with an *ought*, or an *ought not*. [...] For as this *ought*, or *ought not* expresses some new relation or affirmation, 'tis necessary that it shou'd be observ'd and explain'd; and at the same time that a reason should be given, for what seems altogether inconceivable, how this new relation can be a deduction from others, which are entirely different from it. ((1738) 1969: 521)

In order to overcome this problem one must provide an incontrovertible normative principle that can mediate the is- and the ought-statements. In his attempt to do so, however, Jonas falls short.

His answer to Hume is twofold. The first aspect relies for its persuasive force on the concept, discussed in Chapter 2, that we called the *nisus* of Being. Jonas suggests that the biosphere has an instrumental value in systemically sustaining the accomplishments of the *nisus* of Being – namely, life itself – as becomes clear in the following quotation:

> Great is the power of tigers and elephants, greater that of termites and locusts, greater still that of bacteria and viruses. But it is blind and unfree, although driven by purpose; and it finds its natural boundary in the counterplay of all the other forces which carry on the natural purpose just as blindly and choicelessly and in the process hold the manifold whole in symbiotic equilibrium. It can be said that here the natural purpose is administered severely but well, that is, *the intrinsic task of being fulfils itself automatically*. (IR: 129; emphasis added)

Jonas here refers to a 'task' and 'purpose' of Being, thereby lending the notion of *nisus* a fully teleological aspect that, properly understood as a tendency, it does not possess. We can therefore only make sense of the quotation by assuming that he is speaking figuratively. But the key point to grasp is that life as such, preserved in being by the interrelationships of the biosphere, is an accomplishment – one unforeseen and by no means inevitable – of the *nisus* of Being.

What has this to do with the refutation of the is–ought gap? It is intended as the ultimate refutation of the doctrine that nature does not care one way or the other, that what-is says nothing as to what-should-be. For if Jonas is correct, in its *nisus* we bear witness to a 'preference' in Being for the manifestation of life and inwardness over their remaining silent, and, if so, it is here that the deepest axiological vein can be located. In one of the pivotal passages of *The Imperative*, Jonas argues that '[i]n purposiveness as such [...] we can see a fundamental self-affirmation of being, which posits it *absolutely* as the better over against nonbeing. In every purpose being declares itself for itself and against nothingness' (*IR*: 81). Elsewhere Jonas articulates the idea as follows: '[L]ife says "yes" to itself [...] which ever reasserts the value of Being against its lapsing into nothingness' (*MM*: 91). In either formulation the crucial point is that Being as such is not axiologically neutral. On the contrary, having given rise to life – which 'is its own purpose, i.e., an end actively willing itself and pursuing itself' – Being, that which *is*, reveals its preference as to what *ought* to be (*MM*: 173).

Assuming one finds Jonas's speculative arguments for the *nisus* of Being persuasive, we have here a powerful axiological challenge to modern nihilism. Nevertheless, by itself it is insufficient for the express purpose of overcoming the is–ought gap. As Jonas himself concedes, it does not logically follow from the fact that Being says 'yes' to itself that it is objectively good; '[I]t can always be doubted whether this whole toilsome and terrible drama is worth the trouble' (*IR*: 49). In other words, even if we accept that each living being has a good of its own and possesses the status of an end-in-itself, and even if we accept that the realization of such beings is the result of a tendency immanent in Being itself, this cannot be self-evidently described as *objectively* good.

To this end, and constituting the second step in his response to Hume, Jonas rather disappointingly opts for an argument from intuition: he claims as 'axiomatic' and grasped 'with intuitive certainty' that 'the mere capacity to have any purposes at all [is] a good-in-itself' (*IR*: 80). Obviously the veracity of this intuition, even if universally shared, cannot be simply assumed – hence the scorn Jonas drew from his German critics. In Jonas's defence he does offer a justification of sorts: that the denial of this axiological axiom is paradoxical, since it would betray a value-preference for the non-existence of values. Jonas wonders whether this makes his claim an analytical statement, but admits that he is 'not certain' that it does, only that 'there is plainly no going back behind it for something more basic to underpin it' (*IR*: 80).[3]

Theresa Morris has sought to defend Jonas's second argument against Hume by suggesting that although valuation 'is subjectively based in each individual being's desire for existence, it is objectively present in that it is a universal value' (2013: 117). By this Morris appears to mean that the universal presence of subjective valuation (i.e. in all living beings) makes that value objective in that it transcends the individual subject. Unfortunately this interpretation trades on a conflation of the very different terms 'universal' and 'objective': the fact that subjective value is universally present in living beings makes the value *itself* no more objective than if it were to reside in just one individual. For the distinction between objective and subjective is of an entirely different order to that of the universal and the particular: the latter refers to the *scope* of value, whereas the former pertains to the *status* of the value. To be objective a value would have to exist independently of any valuing being – a good-in-itself – and it is this that Jonas can only resort to demonstrating through intuition.

After what Gerald McKenny calls Jonas's 'herculean labors' to identify an objective value, the fact that Jonas concludes his axiological investigations with an argument from intuition is somewhat anticlimactic (1997: 62). Although we are able to describe the individual organism as an end-in-itself courtesy of its *teloi*, and possibly even life as such as an end-in-itself based on the *nisus* of Being, we cannot prove that either is really, objectively good: the sceptic will always note in response that these are only subjective facts, and that no binding 'ought' can be derived therefrom. To his credit, Jonas subsequently admitted as much, noting that '[t]he validity of such intuition can, however, be debated; indeed, any individual can deny having it' (*MM*: 107). This is where Jonas's critics are correct: he fails at the last to defeat nihilism by demonstrating the existence of objective values.

IV Moral traditions

There is, however, a way to account for the force of the intuition Jonas cites. Although this alternative concedes any claim to objectivity, it nevertheless allows us to shore up his theory. Lawrence Vogel points towards its basic orientation, suggesting a 'Humean story that would build on feelings', which might not have 'the systematic force of Jonas' cosmic deontology, but it may be more concrete and genuinely persuasive' (1995: 38). In this vein, Jonas himself conceded, after writing *The Imperative*, that a 'combination of biologism and subjectivism (closely related to historical relativism) cannot really be refuted' (*MM*: 108).

Following this line of thought, we can explain *why* the existence of value intuitively strikes us as objectively valuable – even if it is not so in fact. The answer, we tentatively suggest, is that humans are not only organisms but also symbolic beings, and as the latter we inherit particular historical traditions. In the case of most readers of this book, the tradition in question will be one that we might call 'the West', principally formed from the fusion of Greco-Roman and Judeo-Christian culture. This is not to say, of course, either that this tradition owes nothing to any other, or that other cultural currents have not run alongside it throughout the history of the West itself. Clearly both of these statements are correct. Yet there *is* a recognizable Western tradition, rooted in Athens and Jerusalem, that has acted – and continues to act – as the guiding cultural force of Europe and the countries that owe their existence to that continent. By shedding light on the significance of this historical inheritance, we may advocate a 'soft' relativistic defence of Jonas's claim for the objective value of the existence of subjective values. Although this axiological relativism is not, one suspects, a position Jonas would have happily endorsed, it nevertheless allows us to resolve the present issue consistently with his wider thought at the same time as escaping the pure arbitrariness of mere subjectivism.

We saw in Chapter 3 that human beings not only have a world shaped by our organismic being but also a cultural world courtesy of our symbolic being, which extends into and intertwines with the former. Although non-human life expresses itself through movements, sounds and biological functions, a good number of which are intelligible to us, in human life 'this whole natural groundwork is overlaid with system upon system of invented, constructed, and freely manipulated expressions and symbols, culminating in speech and imagery' (*PE*: 249). As such, although bodily expression remains one way in which humans communicate, language is our foremost means of understanding one another. This is particularly true of understanding the record of the past, which is itself evidently not present: 'The word is [...] the eminently "historical" above the substructure of the ever-repeated themes of the species. Through the agency of the word history produces itself; in its medium, it expresses itself; with its record, historical understanding has to deal first and last' (*PE*: 257).

In this way Jonas seeks to explain how language allows for interpersonal and trans-historic understanding, at the same time as accounting for the 'subhistoric' and pre-linguistic 'biological dimension which we tacitly presuppose' in understanding all life (*PE*: 253). Although we cannot always be sure of having arrived at an accurate – let alone complete – understanding of other human beings, *as* human beings we nevertheless share a common biological-symbolic ground which allows for understanding to take place. It must be said, however, that in stressing the possibility of shared understanding across history and culture, Jonas arguably overstates the extent to which this occurs. That is to say, he does not offer a whole account of our hermeneutic situation, which would also include what is *particular* to historical traditions and inaccessible to our organismic sub-historic understanding. One such case, it will be argued, is that of value judgements. To begin to make sense of this claim we shall turn to Hans-Georg Gadamer – another of Heidegger's students, and a contemporary of Jonas's – for further elucidation of our historicity.

For Heidegger himself, our existing in a historical moment constituted the 'fore-structure' of *Dasein*'s existence, meaning its hermeneutic situation: that, initially at least, we cannot understand the world except according to the context in which we happen to exist (2010a: 146–7). Even if this picture is exaggerated – as Jonas shows with reference to the understanding of life inherent to corporeal being – it is nevertheless instructive regarding the cultural world we inhabit as symbolic beings. More so even than Heidegger, it is Gadamer who brings this line of thinking to fruition. Central to Gadamer's philosophical hermeneutics are the concepts of tradition and horizon, which together illuminate our historicity. The general hermeneutic method begins by looking beyond our immediate circumstances to the historical and cultural context in which understanding is situated. To do this is to acquire what Gadamer calls 'historical consciousness' of living within a tradition – where 'tradition' denotes not something local or institutional, but rather the framework of intelligibility that influences how phenomena show up as meaningful (2004: 303). This temporal and cross-cultural perspective is the condition of possibility for historical and social scientific enquiry, but a philosophical hermeneutics cannot stop there. For what is essential, yet goes unacknowledged in historical consciousness, is that historical and cross-cultural understanding not only occur *from* a particular time and place but are also *constituted by* that time and place. Recognition of this situatedness, which Gadamer calls 'historically effected consciousness', transforms our self-understanding (2004: 301). We then realize that to understand historically or anthropologically is not to lever consciousness out of one historical–cultural tradition and into another. It is, on the contrary, to recognize that all such understanding must in fact take place from a given standpoint.

Historically effected consciousness is not hermetically sealed in a place and time, however: clearly, it is at least possible to revise our scheme of understanding on the basis of an encounter with another. It is because of this openness that Gadamer characterizes consciousness as a 'horizon', and the bringing together of two different historically or culturally situated beings a 'fusion of horizons' (2004: 301, 305). This notion can be better understood by considering a real-life example. Let us imagine a Western anthropologist undertaking ethnographic research of a Pacific island society. Philosophical hermeneutics can help us see how the anthropologist understands the behaviour of the islanders to whose culture they do not belong. Jürgen Habermas describes the central structures of difference that must be breached in order for understanding to take place: 'Each of the partners between whom communication must be established [...] lives within a horizon. [...] This is true both for the vertical plane, on which we overcome a historical distance through understanding, as well as for the process of understanding on a horizontal plane, which mediates a linguistic difference that is geographical or cultural' (1988: 151). The example of the anthropologist and islanders is a horizontal – that is, geographic and cultural – rather than a vertical, or historical, act of interpretation. But either makes the encounter a hermeneutic challenge: the greater the historical or cultural distance, the greater the obstacle to achieving a fusion of horizons.

Recognition of the possibility of reaching an understanding rescues Gadamer's hermeneutics from a hard relativism. For the fact that a fusion of horizons *can* take place entails that both parties have a shared ground that allows for horizontal and vertical differences. But what is this shared ground? Gadamer argues that it is language. Inheriting a particular language that governs much of our lives, inside and out, in thought and in action, we not only possess language, but are at the same time possessed by it. Gadamer goes so far as to claim that language 'operates in all understanding' (1976: 29), and is '*the universal medium in which understanding occurs*' (2004: 390). Although this is not entirely true, as Jonas showed with regard to pre-linguistic corporeal understanding, it certainly is the case that our symbolic being accounts for the majority of what can be expressed and understood. Just as what we do cannot be understood without reference to the body that we are, what we say and think is made possible by the language we are delivered over to. Together these constitute the horizon of our understanding, and therefore the conjoining of Gadamer and Jonas's hermeneutics allows us to acquire a balanced picture of our historicity.

Now, what has any of this to do with the question of the goodness of Being, which motivated our turn to hermeneutics? The answer is that an understanding of our historicity allows us to explain why the existence of value is liable to strike us as objectively valuable. The connecting thread is that our adoption of a historico-linguistic tradition informs what we take to be valuable; in other words, our horizon of understanding is also an *axiological* horizon. This is compatible with Jonas's philosophical anthropology, which stresses the capacity for judgement as a formal aspect of our symbolically mediated existence, but does not presuppose a universal axiological content. Specific axiological judgements are instead accounted for by the tradition that we inherit – even if the capacity for judgement itself belongs to *that which* inherits.

Here we may draw on the notion of an ethical tradition lying at the heart of Alasdair MacIntyre's moral philosophy, which is, by his own admission, heavily indebted to Gadamer (2002: 171). MacIntyre begins by observing that the 'Enlightenment project' of attempting to rationally demonstrate what is good, just and virtuous has not only failed but *had* to fail. The reason why is that, as we have seen, the scientific revolution ultimately led to the discrediting of species essences. This entailed the 'elimination of any notion of essential human nature and with it the abandonment of any notion of a *telos*' towards which ethics, as it had been previously understood, was oriented (2007: 55). Left in its wake is 'a moral scheme composed of two remaining elements whose relationship becomes quite unclear': on the one hand, 'a certain content for morality' belonging to the Greco-Roman-Judeo-Christian tradition, and, on the other hand, 'a certain view of untutored-human-nature-as-it-is' (2007: 55). As the original purpose of ethics was to help realize humanity as it *should* be, our moral norms 'are clearly not going to be such that they could be deduced from true statements about human nature', which was precisely what the Enlightenment moral philosophers had attempted to do – Kant from the perspective of reason, Hume from the passions (2007: 54). All subsequent attempts to philosophically justify the content of our moral tradition – and Kant's has not yet been matched for philosophical brilliance – were doomed to fall victim to the same problem.

What MacIntyre argues with specific regard to ethics we may extend to axiology. Our axiological judgements cannot be rationally demonstrated to hold objectively, but can instead only be accounted for through a combination of sentiment and custom. This is to say that our understanding of objective goodness is grounded in emotion rather than reason, and that its content is largely shaped by the tradition to which we belong. Taking the former element first, Hume was right: whatever we approve or disapprove of cannot be shown to pertain to values objectively correlating to the facts of the matter, but only *our* appraisal of those facts. The is–ought gap has, in other words, proven insurmountable, as demonstrated by the very problem Jonas's philosophy encounters: he is able to rationally demonstrate the existence of subjective values in living beings, but cannot, in the end, demonstrate the objective value of these values, and so appeals to an argument from intuition. Nevertheless, the latter has a strong persuasive force, and this is due to the sentimental appeal that reasons can conjure. Hume was again correct to observe that although 'reason is perfectly inert, and can never either produce or prevent any action', it *can* alert us to something emotionally compelling, leading us to act accordingly ((1738) 1969: 509).

Hume was incorrect, however, to suppose that what 'constitutes virtue or happiness and vice or misery' – that is, a specific account of the good – 'depends on some internal sense or feeling which nature has made universal in the whole species' (1975: 169). On the contrary, our axiological judgements are ultimately informed by our cultural and historical context. Beauty and ugliness are the most obviously relative, as demonstrated by the wide variety of aesthetic standards held across known cultures, but the same is true of ethical value judgements. Although in the contemporary West we consider the suffering of the innocent to be an evil, it is fair to say that our pre-Christian ancestors did not. And because such judgements are ultimately rooted in emotion rather than reason, there is no objectively correct axiological judgement independent *of* tradition. As Gadamer says, 'The real force of morals [...] is based on tradition', this being 'the ground of their validity' (2004: 282).

In our case, as stated, that tradition is a fusion of Greco-Roman and Judeo-Christian ethics, and it is on *this* basis that Jonas's claim for the intuitive certainty of the objective value of values is persuasive. At the bottom of Aristotle's philosophy is the notion that 'the living, having soul, is thereby better than the lifeless which has none, and being is better than not being' (1984a: 731b), while even the book of Genesis states that when God created the earth and all that lives upon it – plant, animal and human – He saw that it was good (1.4–31). This commitment running through the Western tradition, though locked in an eternal struggle with the gnostic tendency that would deny the goodness of nature, means that the combination of a teleological understanding of life and the demonstration of its axiological dimension is indeed liable to bring with it an intuition of objective goodness.

It should be noted that while our axiological judgements are accounted for by sentiment and custom rather than reason, it is evidently the case that they are not immune to rational contestation. Indeed, we have tacitly demonstrated this ourselves in arguing for a biocentric rather than anthropocentric axiology, in partial dissent from our tradition. How can this be explained? For one thing, as Gadamer observed, traditions are neither isolated nor monolithic, but rather subject to the forces of history and intercultural exchange. Secondly, and crucially, just as reason can alert us to something that makes an emotional appeal of us, so too can it shift or weaken the emotional bonds of a tradition. Indeed, a great part of the history of ethics has consisted of the constructive questioning of tradition: here we need only name the Buddha, Jesus, Socrates, or (exemplifying *de*structive questioning) Nietzsche. Even if it is socially and psychologically hard to actually live by such a deviation – it being no coincidence that two of these figures were put to death, and the remaining two self-imposed exiles – it is very much possible to intellectually do so. The reason for this is, as Habermas notes, that questioning the tradition one belongs to breaks its 'quasi-natural' status: even if its content is subsequently accepted, one's relation to it is altered through the act of contestation (1988: 168). Here the freedom of our reason and the inheritance of a tradition are in productive tension, allowing for the latter to change.

Since we *can* reject the aspect of our tradition that explains the intuitive force of Jonas's axiological axiom, the question then is whether we *should* do so. The answer is surely 'no', simply because believing that the existence of life is good and that Being is better than non-Being fundamentally informs what Jonas calls 'the world in which I as a moral being am able to breathe' (*OFL*: 13). That is to say, it determines the very core of who we are, the centre of our self-understanding, and were we to reject it and instead accept the reverse then we would truly fall victim to the gnostic forces of nihilism. For those seeking a proof this justification may well disappoint. But on the question of objective values there is no proof, only affirmation or renunciation. On this basis – which is rather different to the one Jonas envisaged, but still consistent with his wider thought – we can justify a collective 'No' to nihilism and 'Yes' to life and Being, which will serve to underpin Jonas's great contribution to ethics: the imperative of responsibility.

Before looking to Jonas's ethics, however, let us sum up our findings thus far. Just as Jonas's philosophy of life seeks to overcome modern gnostic dualism, so too does his axiology seek to counter modern gnostic nihilism. We have seen that all living beings are immanently teleological both in their activity and self-organization, and that even Being itself may have a tendency towards life which we have here called its *nisus*. These findings represent a radical break with modern ontology, with similar consequences in axiology, as demonstrated in the present chapter. Courtesy of their immanent-teleological constitution, all organisms have the capacity to value both instrumentally and intrinsically. And, as the organism's organization and much of its behaviour are oriented towards its continued existence, each living being earns the status of an end-in-itself.

We must admit that Jonas's axiology is less successful than his ontology, however. That the organism has a good of its own follows as a matter of course from its teleological basis. But whether the existence of subjective value is itself objectively valuable cannot be rationally demonstrated, the is–ought gap proving insurmountable. Nevertheless, the intuitive force of Jonas's claim that it *is* objectively valuable can be accounted for on the basis of our moral tradition. Critical to the latter is the belief that living is better than non-living nature, purpose than purposelessness, Being than non-Being. And we are sure as a result that the existence of value is itself objectively valuable. Even if this is only, at the end, a relative form of value, it represents a deeper form of relativism than is typically encountered, and surpasses the mere subjectivism that Jonas saw as exemplary of modern nihilism. On this basis we turn now to the question of ethics.

5

New dimensions of responsibility

I Ethics, old and new

In Chapter 4 we saw that Jonas's axiology amounts to an anti-gnostic revaluation of values. Far from a value-free domain, we showed that life is the condition of possibility for there being value at all, and that the movement towards life in Being may be taken to constitute a cosmic affirmation of the existence of value. Although we were not able to categorically prove that this fact is itself objectively valuable, we were nevertheless able to justify its intuitive appeal on 'deep' historically relativistic grounds.

Armed with this axiological insight we shall follow Jonas onto ethical terrain. His reason for moving from the former to the latter pertains to the analysis of modern gnosticism recounted in Chapter 1. Recall the quasi-secularized eschatology that Bacon and Descartes had propounded: that the technological mastery of nature, both human and non-human, was the path to utopia. Now, this particular eschaton could, logically speaking, have been conceived of at any point from the evolutionary arrival of the *animal symbolicum* onward, which had available to it the first tools. But even if it had been thought of little would have been practically achieved, as only modernity's materialist revolution and devaluation of nature made radical technological acceleration possible. Of course, the techno-utopian vision eventually found a practical expression in the industrial revolution of the eighteenth and nineteenth centuries, and, more recently, in the development of biotechnological means of modifying living beings. Yet neither has delivered on its promise of benevolent abundance, as both pose problems of a scale and kind never before encountered. Hence we require an ethic that rises to the challenge, countering the giddy utopianism of modern gnosticism with a sober demonstration of the global and temporal reach of our *responsibilities*.

Since modern gnosticism construes both the human body and the natural world as objects of technological manipulation, Jonas's moral response falls within the sub-disciplines of bioethics and environmental ethics – both of which he had a part in defining. One of the debates central to environmental philosophy is whether the ecological crisis requires new ethical principles by which to act individually and collectively, or whether the necessary ideas are already present – albeit supressed – in traditional moral thought. An equivalent question is largely absent from Anglophone bioethics, meaning that Jonas's search for a new ethic appropriate to the technological age is best situated in the former area of practical philosophy. We shall, therefore, begin with this question and Jonas's answer to it.

Jonas begins by exposing the limitations of previous ethical systems, arguing that all share four basic problems. First, he says, the integrity of the biosphere 'did not constitute a sphere of authentic ethical significance', as human action could not affect its equilibrium (*IR*: 4). Second, our dealings with individual plants and animals were thought to not concern morality – hence Jonas claims that 'all traditional ethics is anthropocentric' (*IR*: 4). Third, '[T]he entity "man" and his basic condition was considered constant in essence and not itself an object of reshaping *techne*' (*IR*: 4).[1] Fourth, and finally, 'The good and evil about which action had to care lay close to the act, either in the praxis itself or in its immediate reach [...]. Ethics accordingly was of the here and now' (*IR*: 4–5; emphasis removed).

The clearest example of an ethic sharing all four characteristics is the biblical exhortation to 'Love thy neighbour', which may well serve to epitomize the moral core of the Western tradition. Evidently, it refers only to human beings in temporal and spatial proximity, and for most of human history such an injunction would have been largely adequate as a means of regulating human behaviour. Although pre-modern societies certainly made greater inroads into the natural world than they were (presumably) aware of at the time, it is true to say that overall the symbiotic balance went undisturbed.[2] The human condition could, moreover, have been reasonably thought of as eternal, while the effects of any possible human action were far from global. The sole exception to the appropriateness of the moral edict in question concerns Christianity's ethical treatment of non-human life, to which Jonas apportions significant historical blame for our ecological predicament.

As we have seen, Jonas holds that the *direct* origin of the ecological crisis lies in modernity's reconceptualization of nature as matter, a theoretical revolution that gradually stripped Being of teleology and hierarchical significance. However, he claims that this was only possible on the basis of the West's broader gnostic tendency, of which Christian metaphysics is one component. The Judeo-Christian creation myth, in contrast to Greco-Roman cosmology, construes the earth as teleological in the transcendent sense – that is, as an artefact. 'Thus the idea of a mindless or "blind" nature which yet behaves lawfully – that is, which keeps an intelligible order without being intelligent – had become metaphysically possible' (*OF* I: 17). The exception to this is, of course, humanity, which is uniquely made in the image of God. With this exalted position comes a rightful dominion over nature, to use it to serve our ends as we see fit, and in the book of Genesis we read that it was Adam, rather than God, who enjoyed the symbolic power of naming every living creature, representing 'the first step towards man's coming mastery over nature' (*OF* V: 36). For these reasons Jonas concludes that the Judeo-Christian world view is at the bottom of the West's 'ruthless anthropocentrism' (*IR*: 45).

Now, Jonas's attribution of blame to Genesis is somewhat excessive. We already noted in Chapter 4 that in the creation myth God proclaims the newly created earth to be good in and of itself, that is, prior to serving any human needs. Moreover, as J. Baird Callicott has observed, the Bible sanctions both despotism over nature *and* a countervailing ethic of stewardship, according to which human beings hold the earth as a trust for the sake of future generations. Curiously, these divergent readings are both rooted in the book of Genesis, which Callicott notes is likely woven together from different sources (1989: 138). The priestly version of creation (comprising Gen. 1–2.3) contains the key textual support for the despotic reading, while the older Yahwist creation myth (2.4–4.26) better supports a stewardship interpretation: the command to 'dress' and 'keep' the Garden, in particular, implies cultivation and a relation of care (Gen. 2.15). Jonas himself acknowledges this on occasion, speaking in a separate essay only of the limitations of '*most* former ethical systems, religious and secular' (*TSE*: 894; emphasis added). In line with this weaker claim it can be plausibly argued that the central components of the Western moral tradition – Greco-Roman virtue, Judaism, Catholicism and Protestantism – have generally conformed to the four characteristics of traditional ethics Jonas gives. The foremost exception is the stewardship ethic, which accordingly serves as the model for a new ethic of responsibility (*IR*: 8). Regarding the foundation of a new stewardship ethic, however, Jonas is adamant that it can only be secular, as 'religion in eclipse cannot relieve ethics of its task' (*IR*: 23). But this is, he suggests, a wholly new challenge for philosophy. Let us see why.

II The temporal horizon

The need for a new secular, philosophical ethic can be best justified with reference to the temporal horizon of power radically expanded by technological civilization. Traditional philosophical ethics – whether utilitarian, Kantian or Aristotelian – can, to be sure, solve the problem of anthropocentrism that Jonas identifies in Judeo-Christianity, and thereby take on a global environmental dimension. But the truly novel threat, which none of these theories can accommodate, pertains to the future, since through either the biotechnological reshaping of humanity or through a global calamity such as nuclear war or catastrophic climate change, future generations might not come to exist. Why then *should* human beings continue to be? Would anything essential really be lost in the event of our extinction or radical transformation, and if so, are we obliged to preserve it? These questions, which in earlier epochs must have seemed strictly hypothetical, now very much demand answers. And yet no traditional moral philosophy can provide any.

This failure most obviously applies to virtue ethics. Jonas begins *The Imperative of Responsibility* with a reading of the Chorus of Sophocles' *Antigone*, as an example of the ancient Greek ambivalence towards nature (*IR*: 2–4). According to Sophocles, nature is to be respected as a force in its own right, almost as a powerful foe. Clearly, this attitude is no longer appropriate, and perhaps no longer even possible since we have so effectively subdued nature. Regardless, even if we adopted an environmentalist version of virtue ethics – an ethics of respect for nature, say – it would not be able to accommodate considerations of future generations without simply assuming their existence, since the cultivation of good character requires that there *be* humans in the first place.[3]

The same basic limitation affects rights theory. On most models, rights are derived from an individual's justified and inalienable claim to something, life being 'the most fundamental of all rights' on which all others rest (*RD*: 31). Whether morally or merely legally grounded, rights entail the obligation of others to uphold them in practice: hence we might suppose that future generations have a right to life, entailing duties for us to procreate and leave behind a habitable world.[4] But, as Jonas correctly points out, only an entity that has claims can make them of us, which first of all means that the entity must *exist*. A member of a future generation, as inexistent, self-evidently 'makes no demands and can therefore not suffer violation of its rights' (*IR*: 39).[5] It is clear, therefore, that the concept of moral rights cannot establish obligations to future generations without simply assuming their existence, which is precisely what is newly endangered.

This problem applies equally to Kantian and consequentialist attempts to account for the ethical treatment of future generations. Taking utilitarianism as our representative of the latter, its object is, broadly speaking, to maximize happiness for the greatest number of people. Of course, this requires that people are alive in order that their happiness can be taken into consideration, and again, what the theory cannot do is explain why there ought to be people in the first place. The utilitarian might suggest that it is better for people to exist than for them not to, as it is better for some happiness to exist than none at all. But when we ask who exactly it is better for, the utilitarian can only respond: for those assumed to exist. For this reason consequentialism, like a rights-based approach, represents an insufficient ethical framework in light of our current predicament.

Finally there is Kant's theory, which at its most elementary states that I must act in accordance with the moral law holding for each rational being. This essentially means that the principle guiding my action has to be binding on the basis of its universalizability, without reference to ends that would render it applicable only in certain cases. Hence an imperative that takes the form 'if x then y' is merely hypothetical, while an imperative that holds for all moral agents under all circumstances is categorical. For this reason the principal formulation of Kant's categorical imperative is as follows: 'Act only according to that maxim whereby you can at the same time will that it should become a universal law' (1993: 30). This is a purely formal version, however, and Kant provides two further formulations of the categorical imperative – one moral, the other political – that he held to be synonymous.[6] The second formulation requires us to '[a]ct in such a way that you treat humanity, whether in your own person or in the person of another, always at the same time as an end and never simply as a means' (1993: 36). And the third formulation runs: 'A rational being must always regard himself as legislator in a kingdom of ends rendered possible by freedom of the will, whether as member or as sovereign' (1993: 40).

Jonas argues that all three formulations of the categorical imperative – the formulae of Universal Law, Humanity and the Kingdom of Ends, respectively – once more pertain only to already existing beings and those that we assume will exist: '*Given* the existence of a community of human agents […], the action must be such that it can without self-contradiction be imagined as a general practice *of that community*' (*IR*: 11; emphasis added). The reason for this restricted scope of application is that '[j]ust as I can will my own end, I can will that of humanity. Without falling into contradiction with myself, I can prefer a short fireworks display of the most extreme "self-fulfilment", for myself or for the world, to the boredom of an endless continuation in mediocrity' (*IR*: 11).

Jonas's claim that 'I can will my own end' is admittedly odd since Kant famously argued that suicide violated an 'irremissible', or perfect duty – as opposed to a 'meritorious', or imperfect duty – to oneself (1993: 32). Kant says: 'One sees at once a contradiction in a system of nature whose law would destroy life by means of the very same feeling that acts so as to stimulate the furtherance of life' (1993: 30). Jonas is right, however, to say that Kant's theory allows one to act in a way that compromises the conditions for future generations. We can formulate the maxim of a hedonistic and environmentally reckless course of action as follows: 'I will live a life of consumption that is beyond ecologically sustainable limits for the sake of greater pleasure.' Upon universalization we of course see that the earth's resources would quickly expire, leading to the impossibility of future generations living so indulgently, and perhaps not existing at all. For Kant this would no doubt be an unattractive prospect, but it does not entail a contradiction in conception: adopting such a maxim appears to be logically valid since it does not lead to the destruction of any present-day human beings who are able to act according to a maxim of self-indulgence. Again, it only applies to future generations if we assume their existence, and consequently we cannot on this basis derive a perfect, irremissible duty to live sustainably and so guarantee their existence.

On the other hand, and maintaining the assumption that the first formulation of the categorical imperative binds the actions of a contemporary community, one might address the issue from a proactive angle by asking whether there is a moral obligation to procreate. Indeed, Kant believed that procreation formed part of 'a human being's duties to himself', not in virtue of one's rationality but rather as an end of one's 'animality' (1996: 175). If such a duty could be established, then Kant's deontology would seemingly deal with the problem of future generations' existence, as each generation would be obligated to generate the next. If, for example, an anti-natalist were to formulate the maxim of their action – 'do not procreate' – and universalize it, we see that this would lead to the extinction of humanity. Once more, however, although this is an unappealing prospect, it does not seem to constitute a contradiction in conception, since the eventual extinction of humanity is again compatible with willing the end of that maxim: a refusal to procreate is perfectly compatible with the identical actions of those beings capable of being bound by the moral law.

As such, Jonas is right to say that a duty to ensure the future existence of human beings represents something of an aporia for moral philosophy. This is exactly the problem that Jonas's theory is intended to solve, by conceiving of the preservation of humanity as a general duty from which the individual's duty to act accordingly is derived. At the same time this will, as we shall see, necessitate a global environmental ethic, thus simultaneously correcting the (purported) excessive anthropocentrism of Judeo-Christianity.

III Responsibility for the 'idea of Man'

We now turn to Jonas's solution to the aforementioned problem. Having shown the existing theories to be individually inadequate, Jonas nevertheless draws on aspects of Kant's moral philosophy to develop an ethic that could supplement – not replace – traditional morality. Like Kant, Jonas seeks to give philosophical expression to the Western 'ethics of consciousness' and 'respect for man', yet does so in a distinctive way (*OPW*: 199). He states that 'the intrusion of distant future and global scales into our everyday, mundane decisions is an ethical novum which technology has thrust on us; and the ethical category pre-eminently summoned by this novel fact is: *responsibility*' (*TSE*: 893). Why? Because responsibility is a correlate of free action, and 'the claims on responsibility grow proportionately with the deeds of *power*' (*TSE*: 893). Since we now have the power to dictate not only the condition but the very *existence* of future generations and the biosphere, we require an ethic that can accommodate this change.

Although secular in orientation, Jonas explicitly connects his new ethic of responsibility back to the stewardship ethic contained in the Yahwist sections of Genesis:

> [I]t comes about that technology [...] installs in man a role which only religion has sometimes assigned to him: that of steward or guardian of creation. By enhancing his might to the point where it becomes palpably dangerous to the total scheme of things, technology extends man's responsibility to the future of life on earth, now exposed to, and defenceless against, the abuse of that might. Environmental ethics [...] is the expression of this unprecedented widening of our responsibility. (*TSE*: 894–5)

Seeking to show that this instance of responsibility can be accounted for by reason and phenomenological reflection not only strengthens Jonas's theory but also protects against instances where one is mistaken about having responsibility for others. Three criteria in particular must hold for objective responsibility: *moral agency* on the part of the subject, *moral considerability* on the part of the object, and the subject having *power* over the object. We shall look at each in detail.

Perhaps the most basic criterion for responsibility is moral agency. In everyday life we ascribe responsibility to individual persons: we would say, for example, that if I ran through an antiques shop and smashed a priceless vase, then I would be responsible for this in two ways. Firstly I am *causally* responsible in that the vase smashed because of my actions alone and, secondly, I am *morally* responsible for my reckless behaviour since as a mentally competent adult I should know to act with care around valuable objects, yet failed to do so. Here harm is not done directly to a morally considerable being, but only indirectly to the shopkeeper via the destroyed antique (and perhaps to ourselves, in failing to act virtuously). It is not the first sort, causal responsibility, that we are concerned with here, but the latter, moral responsibility.

Clearly, most beings lack moral responsibility even if they can be said to have causal responsibility for events. Geological phenomena, plants and animals all have causal power – and in some cases a power far greater than any human's – but this alone is insufficient for being morally responsible: while I can be held accountable for recklessly smashing antiques, the bull in the china shop cannot. Only *moral agents*, those with the power to will a course of action, are answerable for their deeds. Adult humans alone are free in this way, a freedom made possible by our symbolically mediated relation to our own bodies that involves the ability to objectify, and reflect on, our own actions and motives.[7] Thus we are morally responsible not only for the actions we do undertake but also for those we ought to have undertaken but failed to, and in both cases this can be true not only of individuals but also of institutions and other such collectives.

As stated, Jonas's theory of responsibility is not merely formal, pertaining to *who* is morally responsible. It is primarily a theory of *what* we are responsible for over and above our own actions: a 'substantive, goal-committed concept of responsibility' for '[t]he well-being, the interest, the fate of others, [wherever this] has, by circumstance or agreement, come under my care' (IR: 93). This sort of responsible care manifests in a variety of ways in everyday life – something Heidegger's account of concern (*Fürsorge*) failed to show – but in moral and political philosophy it exists primarily in the social contract theories of Hobbes, Locke, Rousseau and so on, and, as we have seen, in the Judeo-Christian tradition of stewardship. Stewardship is an example of circumstantial responsibility, God having supposedly placed humanity in charge of the natural world, while the social contract is evidently responsibility by agreement (or, in its weaker forms, tacit consent). If the notion has played scarcely a greater role in Western philosophy than this, it is because, as Jonas says, this form of responsibility tracks the *power* of the moral agent over the morally considerable being: a power which was historically confined only to the immediate. Power, our second criterion, is essential to the phenomenon because to have responsibility *for* something entails that it comes under our sphere of influence; it must be something that we can, at least in principle, act upon. Where such causal power is lacking, moral responsibility is a logical impossibility: one could not, for example, be responsible for the condition of life on distant planets (if indeed it exists), since it lies beyond our power to affect.

Of course, power over another being does not alone entail responsibility for it. For instance, on a whim I could destroy all the material goods in my possession and yet have done nothing wrong, as the objects themselves are not morally considerable. Power only becomes a matter for ethics in those cases where it directly or indirectly concerns a morally considerable being: it is this that ultimately grounds 'the ought-to-do of the subject' (*IR*: 93). Our third criteria, then, is *moral considerability*. We have already covered Jonas's argument for the existence of morally considerable beings in Chapter 4, arguing that the immanent teleology of living beings meant that these alone were ends-in-themselves in possession of a subjective good. Unfortunately Jonas's argument for the objective value of that value was ultimately found wanting, and therefore any responsibility for these beings cannot be incontrovertibly proven. Nevertheless, we argued that according to the moral tradition to which we belong, the existence of subjective goods could be understood as objectively good, and on these relativist grounds we may transcend pure subjectivism to reach a responsibility for others that exists 'over against the will' (*IR*: 84).

Responsibility of the sort Jonas is concerned with is not only a matter of objective criteria being met, however. Central to the phenomenon is that it is also *felt*. For this reason Jonas seeks to phenomenologically account for the way in which the morally considerable being makes a demand upon the moral agent, which he characterizes as being 'called to its care' (*IR*: 93). In doing so Jonas surely refers to Heidegger's analysis of conscience as 'the call of care', although for Heidegger this meant only that *Dasein* was called out of its lostness in the they and back to its ownmost-potentiality-of-being (2010a: 277). Jonas, however, has almost the very opposite in mind, as here the call of care is explicitly a matter of ethics. It occurs when the conditions of responsibility meet in our encroaching on a *vulnerable* being, the 'perishable *qua* perishable', which, having ends and thus a subjective good, is open to harm (*IR*: 87). Consider, by contrast, a being that was both absolutely good and invulnerable: a god, say. Despite its goodness, such a being cannot be an object of responsibility since we *objectively* have no power over it and its invulnerability makes no *subjective* demand of us, the two sides being complementary. The true call of responsibility depends on bearing witness to vulnerability and its possible protection through our person. Hence Jonas modifies Kant's dictum 'ought implies can' to instead read: 'You ought *because* you can' (*IR*: 128; emphasis added).

In stressing both the cognitive and non-cognitive aspects of responsibility, Jonas goes a step beyond Kant, who had argued that 'moral worth depends […] not on the realization of the object of the action, but merely on the principle of volition according to which, *without regards to any objects of the faculty of desire*, the action has been done' (1993: 13; emphasis added). To be sure, emotion plays a supplementary role in Kant's ethics, but only through our 'reverence' for the 'moral law within' ((1997) 2015: 129; emphasis removed). What this means in practice is that the worth of our action depends not on our commitment to the being to which the moral law applies, but rather on our commitment to the moral law *itself*. Evidently this follows from the formalism of Kant's ethics, yet it cannot but feel inadequate as an account of moral worth when we consider the case of being called to care for the other.

Jonas's approach is intended to bridge reason and emotion, Kant's ethics and Hume's, by arguing that in the case of responsibility 'the two sides are mutually complementary and both are integral' (*IR*: 85). Thus he writes: '[N]ot the moral law motivates moral action but the appeal of a possible good-in-itself in the world, which confronts my will and demands to be heard – *in accordance with* the moral law' (*IR*: 85). For when it is demonstrated that we have a power over a morally considerable being – thereby fulfilling the above criteria – the accompanying feeling is the motivating force to *take* responsibility for it. Conversely, a misplaced feeling of responsibility can be diminished by the rational realization that one is not, in fact, responsible for that being, should any of the necessary criteria not hold. As such, we may account for responsibility as a moral phenomenon ultimately grounded in sentiment, but which is nevertheless responsive to reasons insofar as they pertain to the criteria outlined.

As the foregoing suggests, Jonas's theory entails that the 'precarious, vulnerable, and revocable character, the peculiar mode of transience, of all *life*, […] makes it alone a proper object of "caring"' (*IR*: 98). Although one might intrinsically value certain inanimate beings, such as masterpieces of art and architecture, one can only care for them (in the ethical sense) for some other purpose which has its origin *in* a living being: that beauty elevates the soul, for example. Thus only life is a proper object of responsibility. More will be said specifically about responsibility for non-human life later in the chapter; first of all, we must look to one instance of responsibility in particular that stands out in our experience of the phenomenon.

Jonas observes that the 'timeless archetype of all responsibility' is the parent's responsibility for the newborn, which is 'so spontaneous that it needs no invoking the moral law, [and] is the primordial human case of the coincidence of objective responsibility and the subjective feeling of the same' (*IR*: 130, 90). This responsibility is clearly not contractual, but rather circumstantial, the child having been brought into being through sexual reproduction. This fact alone does not point towards its uniqueness, however – after all, through creative acts we are causally responsible for the existence of many new beings: artworks, buildings, laws and so on. Of course, the newborn, as living, is also a morally considerable being, but even this fails to capture the qualitative difference between responsibility for the infant and any other instance. Note that, as Jonas says, no invocation of the moral law can fully account for its subjective force in this instance. When confronted with a newborn baby its 'mere breathing uncontradictably addresses an ought to the world around, namely, to take care of him. Look and you know' (*IR*: 131). Jonas's point here is phenomenological: the immediate experience of the newborn, shorn of reductive abstractions, is nothing less than perception of an ethical demand: '[T]he plain factual "is" evidently coincides with an "ought" – which does not, therefore, admit for itself the concept of a "mere is" at all' (*IR*: 130).

Phenomenologically speaking this is undoubtedly correct: we do not separately perceive the fact that a newborn child is crying out, helpless, from the demand for protection that it makes on me. Rather, the moral command is *there* in our perception of the child's cry, and only a reprobate could deny it. Logically, however, the distinction between the 'is' and the 'ought' can always be drawn. This move, forcing us on to the terrain of reasons – which are here insufficient – leads us to try to explain, as far as is possible, why we find Jonas's description of the incontrovertible call of responsibility so compelling. Unfortunately the answer is not, as Jonas would like to claim, that we are responsible for the newborn's objective goodness: as we have seen, no moral injunction can be given such an axiological foundation. It is instead because this instance of responsibility is both *the most fundamental ethical experience prescribed by our tradition*, and *follows as closely as possible from our organismic being*.

Since pertaining to our organismic being, it is unsurprising that responsibility for the newborn has antecedents in animal life. In particular, living beings that rear their young in broods realize a form of sociality, of being-with, that is unprecedented in comparison to prior manifestations of that existential structure. As Jonas says, with regard to mammalian life, 'The relation of the mother to her young cannot be compared with anything else in the field of animal emotions and whatever is found elsewhere of protective and tender instincts [–] e.g., also in sex-relations, which of themselves are of much earlier origin in evolution [–] has its roots in the rearing situation' (*OF* IV: 69). Notice that reproduction, the biologically prior fact, is subsequently transformed by the new heights of social existence achieved in rearing: this, Jonas speculates, allows for 'the development of the whole scale of emotions which we comprise under the name of "love"' (*OF* IV: 68–9). Whether or not this latter point stands, it is undoubtedly the case that the intense bond evidenced in mammalian rearing is at the root of moral responsibility for the newborn, since both conform to the same relation between adult and infant. In this way we can understand Jonas's claim that the latter is a responsibility 'instituted by nature' (*IR*: 94). The key difference, of course, is that humans alone can understand intuitive responsibility for their offspring *as* a moral responsibility – a responsibility *for* the end-in-itself of the infant. Since this understanding is permitted by our symbolic being, 'the ontological capacity for responsibility cannot be lost', even though 'psychological openness to it is an historically acquired, vulnerable possession' (*MM*: 106). In other words, history and culture form traditions that bequeath a certain ethical content to which judgements of right and wrong are relative. This entails that although the organismic good of care for the infant *can* be transformed into a moral good, certain traditions do not do so. The practice of exposure, common enough throughout world history, is proof of this. But what matters is that our tradition *does* construe care for infant human life as a moral good; indeed, not only *a* moral good but *the most fundamental* of all such goods, and it is for this reason that we find the sceptic's division of the 'is' and the 'ought' so objectionable here.

Jonas acknowledges that even among cultures that do recognize responsibility for the newborn as a moral obligation, not all have recognized that responsibility as the first of all duties. Nevertheless, he argues that this particular instance of responsibility is worth protecting and promoting due to its unique significance, which is as follows. In terms of duration, responsibility for the newborn is neither momentary nor usually lifelong, but rather dependent on the development of the child, continuing 'until the fulfilment of the immanent-teleological promise of eventual self-sufficiency releases [the parent] from the duty' (*IR*: 131). In other words, as the infant becomes a child, an adolescent, and finally an adult, the parent is gradually released from the total responsibility to which they were initially committed. But this responsibility is not merely tied to organismic development, as is made starkly clear by cases of severe mental disability where the parent's responsibility never expires. What is it, then, that in adulthood relieves the parent of their duty of care? Jonas suggests that it is the achievement of full personhood and the ability to account for oneself: the capacity for moral responsibility. And by extension this means that the *telos* of the parent's responsibility for the infant is the *coming-to-be of another responsible being*.

In this aspect of responsibility for the newborn its objective side takes on a significance that is equal to the phenomenological experience of absolute responsibility. Jonas's argument for the former is as follows. Although all living beings, as morally considerable, can become for us objects of responsibility, the fact that human beings alone are *capable* of taking responsibility sets us *above* the rest of life in the order of moral significance. Not only does this tell us *why* the existence of human beings is our principal responsibility, it also tells us *how* human beings should continue to be: that is, morally responsible. This argument is most explicitly set out by Jonas as follows:

> The appearance of this value [i.e. responsibility] in the world does not simply add another value to the already value-rich landscape of *being* but surpasses all that has gone before with something that generically transcends it. This represents a qualitative intensification of the valuableness of *Being as a whole*, the ultimate object of our responsibility. Thereby [...] the capacity for responsibility as such [...] becomes *its own object* in that having it obligates us to perpetuate *its presence in the world*. (*MM*: 106)

That is to say: humanity is of a qualitatively greater significance than other life thanks to its capacity for morality, a significance that demands humanity's continued existence *as* responsible beings, and to which we are subjectively committed in witnessing the infant *qua* human being. Hence Jonas concludes: 'With every newborn child humanity begins anew, and in that sense also the responsibility for the continuation of mankind' (*IR*: 131).

Now, it will be noted that the object of responsibility that Jonas prizes so highly is not human life per se, but rather our *capacity* for morality. This Jonas happily concedes, stating that 'the possibility of there being responsibility in the world, which is bound to the existence of men, is of all objects of responsibility the first' (*IR*: 99). And again: '[T]*he presence of man in the world* [...] has itself become an *object* of obligation: the obligation namely to ensure the very premise of all obligation, that is, the *foothold* for a moral universe in the physical world' (*IR*: 10). As the latter quotation makes clear, Jonas is not claiming that all human beings are good persons and deserving of continued existence on that basis (nor, it goes without saying, that only good persons are deserving of life). In fact, he suggests that if it were possible to devise a universal calculus of the good and the bad brought about through human history, the latter would probably outweigh the former. Regardless of humanity's dubious moral record, the worth of its existence is accredited by its moral being, and the ever-present *possibility* for moral goodness this represents.

Since the object of his categorical imperative is humanity's capacity for morality, and the ideal of goodness this entails, Jonas suggests that 'we are, strictly speaking, not responsible to the future human individuals but to the *idea* of Man, which is such that it demands the presence of its embodiment in the world' (*IR*: 43).[8] One might then ask how actual human beings and the idea of Man are connected: more specifically, what is it about the former that makes the latter possible?

Jonas's invocation of the idea of Man harks back to the notion of the image that played such a pivotal role in his philosophical anthropology.[9] Indeed, he refers several times in *The Imperative* to the 'image of man' in a manner that suggests the two notions are synonymous (*IR*: 201). This suspicion is borne out by an etymological connection: the image, we recall, captures the *eidos* of the beings it represents, and it is from the Greek *eidos* that the English word 'idea' derives. Recall, moreover, that for Jonas the *eidos* is derived from perception of formally alike beings, and also that it is from our perception of the newborn that responsibility for the idea of Man emerges. The latter connection – between the individual infant before me to which I am responsible and the idea of Man as a responsible being that constitutes the *telos* of my responsibility – is evidently only available to a being that is open to and lives through the *eidos*, that is, the *animal symbolicum*. It is, essentially, a manifestation of our transcendence of the strictly physical world into a world of intertwining beings and *eidoi*. No doubt, in each individual instance of responsibility for the infant none of this is present in the mind of the responsible agent. Yet it is there nonetheless.

In *The Phenomenon of Life* Jonas's discussion of the image or idea of Man suggests that it is a historically and culturally relative invention of a given society: 'That image is worked out and entertained in the verbal communication of society,' he says (*PL*: 186). And yet in *The Imperative* Jonas also speaks of humanity's '*true* image', which is revealed to us, he says, by the contemporary technological threats to humanity's future (*IR*: 26; emphasis added). This means, therefore, that although the *eidos* of the human being takes on a symbolic meaning for each human culture, there is nevertheless one true meaning that goes by the name of *the* idea of Man. The grounds of our epistemic privilege in having perceived humanity's true image are not in themselves a privilege: namely, our existing in the shadow of self-created threats to our future. Exactly how this gives rise to knowledge of humanity's true image will be discussed in

Chapter 6 and then more fully in Chapter 7. For now, we merely note that Jonas's argument for the overriding value of the idea of Man as a moral being represents not just one of many possible versions of the idea but the *true* idea of Man, and it is to this, he suggests, that we are ultimately responsible.

Having argued that this duty is absolute, Jonas summarizes it in a series of purportedly synonymous categorical imperatives, the principal formulation of which is this: 'Act so that the effects of your action are compatible with the permanence of genuine human life [on earth]' (*IR*: 11).[10] Thus his theory is given a Kantian veneer, Jonas's new categorical imperative being intended to supplement the old. Reflection reveals, however, that Jonas's imperative of responsibility only applies when humans already exist. One wonders, therefore, whether it is truly categorical since it appears not to hold in all logically possible situations: for instance, the one in which life failed to realize itself on earth. Jonas is unconcerned by this, admitting as much and claiming that it is nevertheless categorical: 'Groundless itself (*for there could be no commandment to invent such creatures in the first place*), brought about with all the opaque contingency of brute fact, the ontological imperative institutes on its own authority the primordial "cause in the world" to which a mankind once in existence, even if initially by blind chance, is henceforth committed' (*IR*: 100; emphasis added). Accordingly his imperative holds that *if* humanity exists, *then* its existence must be safeguarded on earth. This would mean that Jonas is guilty of exaggerating the exceptionless status of his imperative, but one might defend him by noting that all imperatives share this presupposition yet some are nonetheless conceivably categorical, such as Kant's Formula of Humanity. In other words, *any* moral command must be issued to existing moral agents in order to have purchase; ethics as such is moot without this ontological given. As Jonas's imperative holds in the only world there is, we may say that its status as hypothetical or categorical is for all intents and purposes irrelevant.

With this we may conclude our basic reconstruction of Jonas's theory of responsibility. It is, in short, an argument for the duty to ensure the continued existence of human beings, who, as far as we know, are alone in embodying the transcendent worth of the idea of Man: the capacity for responsibility. And because we are committed to preserving human life on the basis of our moral being, moral beings we must remain. Even if this command is not fully rationally demonstrable – as no moral injunction can be – it is compellingly argued for, drawing on and aligning with the phenomenon of responsibility. In our bearing witness to the newborn in its utmost vulnerability, we are unconditionally called to its care: a care which has its antecedent in immediate organismic being, but which ultimately reaches well beyond this. This is, essentially, Jonas's ultimate riposte to Heidegger, whose analysis of the call of conscience in *Being and Time* was purely formal and devoid of content. Jonas, by contrast, concludes that '[i]n the light of such self-transcending width, it becomes apparent that responsibility as such is nothing else but the moral complement to the ontological constitution of our *temporality*' (IR: 107).

IV Global and intergenerational ethics

We may now turn to the more concrete questions of our duties to future generations and life on earth. These were, we recall, concerns forced upon us by modern technology, and the former of which traditional morality could not fully account for. It must be admitted that Jonas's theory is also limited in its own way, since it permits little fine-grained detail regarding either duty. Specific obligations – the just distribution of resources, say – are not accounted for except in the broadest of terms. This is a consequence of the nature of his theory of responsibility, which is intended to account for our most fundamental duties, thereby acting as a counterpart and orienting principle for complementary theories (Kantian ethics, or a theory of justice, say). For this reason, the following will also remain at a fairly high level of abstraction.

Taking our obligations to future generations first, we may split these into two fundamental categories: obligations pertaining to their *being* and obligations pertaining to their *well*-being. The former is straightforwardly accounted for by Jonas's theory of responsibility since our principal obligation is to the continued existence of human life (objectively, because we have the capacity for morality, grounded phenomenologically in our experience of responsibility for the newborn). This is, after all, the great theoretical achievement of his ethics. The question is then what we owe to future generations regarding their well-being. At the most general level, we may say that following logically from the duty to bring about future generations is the obligation to leave behind a habitable world, which in turn means that a sustainable form of life has to be achieved in the present. As this principally concerns technological and economic systems, rather than individual behaviour, the *means* of achieving this change are largely political, and we shall address how Jonas thinks this should be carried out in Chapter 6.

More specifically, we might ask what resources, institutions and goods we should leave behind. For all moral theories this question necessarily has a degree of uncertainty about it, since we do not know precisely what the material requirements of future generations will be: population booms or slumps as well as technological, social and economic change could all have effects we cannot predict. But we may provide the outlines of an answer by recalling that Jonas's theory demands not only the existence of *human beings* but human beings capable of responsibility:

> [R]especting this transcendent horizon, the intent of the responsibility must be not so much to determine as to enable, that is, to prepare and keep the capacity for itself in those to come intact, never foreclosing the future exercise of responsibility by them. [...] Its highest duty, therefore, is to see that responsibility itself is not stifled, whether from its source within or from constraints without. (*IR*: 107)

Evidently this entails, as a bare minimum, the provision of adequate but sustainable sources of energy and sustenance, shelter and stable political institutions. It is true that Jonas's theory alone cannot tell us in quantitative terms what 'adequate but sustainable' sources of energy and sustenance would amount to: only economists and medical experts can do so. But as a point of principle it can guide the latter. Cultural goods are less obviously accounted for, with the notable exception of education: whether conducted formally or otherwise, the moral component of education is arguably the highest aspect of what we owe to those to come since pertaining to the source of responsibility within. If all sound like obvious aspects of any generational legacy, recall that at present our unsustainable form of life makes their ongoing provision anything but certain.

Remaining at a relatively abstract level of debate, we may switch focus from intergenerational to in*tra*generational ethics, and ask what Jonas's theory of responsibility concretely tells us about the obligations of contemporaries to one another. After all, the ecological crisis threatens the well-being not just of future generations but also of people of the present. Moreover, these harms are not borne equally, nor by those who are best able to cope, nor even by those who are causally responsible for the harms coming about in the first place – any of which would have some claim to being a just distribution. Rather, the harms of climate change and associated ecological degradation will be disproportionately borne by those who are *least* equipped to cope and *least* causally responsible, that is, the citizens of poorer and less industrialized nations.

On this basis it might seem as though a global justice framework would be best suited to account for the wrongs committed against contemporaries and the duties owed to one another: particularly since the inexistence of the vulnerable – the problem Jonas's theory tackles and which no other can – is not here an issue. Nevertheless, the theory of responsibility can contribute to our understanding of this area.

Jonas offers an intragenerational ethic thanks to the nature of responsibility itself. The call of care for the infant, which points towards our futural obligations, may be the first and most forceful example of responsibility, but the phenomenon can be both subjectively felt and rationally extended to one's contemporaries. Subjectively speaking, the feeling of responsibility for other adults is known to all of us. As to its rational justification, Jonas notes that responsibility is, in principle, 'reversible and includes possible reciprocity. Generically, indeed, the reciprocity is always there, insofar as I, who am responsible for someone, am always, by living among men, also someone else's responsibility. This follows from the nonautarky of man', itself a part of the human condition (*IR*: 98). Clearly, there will only very rarely be a total asymmetry of power and vulnerability between adults of the sort exemplified by the parent–child relationship – hence the subjective force of the former is so much weaker than the latter. Nevertheless, socio-economic circumstances can be such that dramatic power imbalances occur (homelessness and destitution, for instance). Ontologically this follows, as Jonas says, from the basic precarity of all life coupled with the social dimension of existence, which for human beings takes the form of communal living. These are then transformed into a relationship of responsibility according to those traditions – and we would be astonished if this did not include *most* moral traditions – that recognize some degree of obligation among kin and neighbour. Of course, ours does compel us to recognize responsibility for the vulnerable good that is in our power to protect; as a consequence, having responsibility for family members, friends, fellow citizens and human beings as such is also appropriate wherever they are endangered and within the reach of our care.

Conversely, our vulnerability means that we too can be objects of responsibility for others, provided the necessary conditions hold. Climate change is one such case, and because of its scale perhaps the most dramatic in human history. With Jonas, we may note firstly that '[s]hared danger surely establishes *reciprocal duties*' among those who are both its cause and subject to its effects, as is true of citizens of the affluent West (*IR*: 236 n.9). Curiously, however, he suggests that the relevant moral demand here is not responsibility, but rather the 'virtues which the trial of the situation may require': in particular, 'courage, resolution, [and] constancy' (*IR*: 236 n.9). The rare invocation of virtue here is welcome, but it is not clear why the presence of the latter should exclude responsibility, rather than assisting in the carrying out of our duties, as Kant would have argued. After all, the logic of the ecological crisis is that we have endangered each other and have the power to alleviate the threat. Since Jonas's criteria are thereby fulfilled, it follows that we the guilty are in fact responsible for protecting one another from the harms of climate change, and that certain virtues will help us do so.

In cases where the shared danger is the result of unilateral action between adults, Jonas suggests that responsibility is indeed the appropriate ethical concept (*IR*: 236 n.9). Of course, this is precisely the case with climate change. Citizens of poorer nations are at risk of ecological harm overwhelmingly because of the actions and behaviour of citizens of industrialized nations.[11] Think, for example, of people in the low-lying Pacific islands at risk of being submerged by rising sea levels: this is predominantly our fault, whereas they have done little, if anything, to bring it about. In a case such as this the emotional side of responsibility is clearly far weaker than that of the parent for the child, but it is nevertheless present, and can be heightened by images or accounts of the threat that traverse the spatial distance. Then it becomes clear that *these* people's lives have been endangered by *our* way of life, and thus the emotive dimension of responsibility may align with the rational.

One might object that Jonas's framing risks condescension: that our having responsibility for the global poor infantilizes the latter, or at least affords less dignity than conceiving of them as candidates for just treatment would. This is unfortunate, if true, since there is no ethnocentrism to be found in Jonas's philosophy: as all humans share the same existential precarity, and fulfil the idea of Man as moral beings, responsibility is always in principle reversible across the globe. In an ideal world, perhaps, no group of adults would have to be an object of responsibility for another. But the present situation, which is a result of historical socio-economic factors, means that we are not equal in terms of the power to affect one another, and this inequality of power entails a responsibility on the part of the affluent: a responsibility to rescue the global poor from the danger in which we have placed them. Jonas's theory of responsibility can thereby not only underpin but also actively orient other accounts of our intergenerational and intragenerational climate obligations. If few specifics can be drawn out, this is simply the price to be paid for such a fundamental theory, which could be complemented by more explicitly rights-based frameworks. These would have to point, however, towards the ultimate object of our obligations: the continued existence of humanity as a moral being.

V Duties to non-human life

We are now in the position to answer a question that has puzzled Jonas scholars until now: Is he a kind of 'Kantian who is decidedly leery of [...] utilitarian reasoning', as Theresa Morris suggests (2013: 167), or is his ethics in fact 'heavily consequentialist' (McKenny 1997: 211)? Due to its emphasis on responsibility as a duty, with the virtues playing only a facilitating role, it seems fair to say that Jonas's ethics is primarily Kantian. However, it departs from Kant's moral philosophy in two key respects. The latter asked only what existing rational beings are obliged to do by the moral law – 'an ever-present order of abstract compatibility' – whereas Jonas's imperative 'extrapolates into a predictable real *future* as the open-ended dimension of our responsibility' (*IR*: 12). His theory is thus teleological in a way that Kant's is not, establishing a duty to the continued realization of a particular kind of being. Secondly, with its accommodation of the emotive aspect of responsibility it has a strong Humean element. We can therefore say that the deontological interpretation is closest to the truth, but not the whole story.

A second interpretative question is harder to answer, however: namely, to what extent is Jonas's ethic of responsibility anthropocentric? Given that he develops a biocentric axiology, built on a teleological philosophy of life, and criticizes the ruthless anthropocentrism of previous ethical systems, it might seem strange that the moral consideration of living beings as such shrinks into the background of Jonas's ethic of responsibility in favour of humanity. Why, after all, should we morally privilege human life if all life forms are ends-in-themselves? Is his doing so an indefensible anthropocentric bias following, perhaps, from stressing the subjective force of responsibility for the infant?

Jonas thinks not, and with good reason. His focus on our obligations to humanity is explained by a conscious attempt to fuse together responsibility for human and non-human life, thus bypassing the anthropocentric–biocentric distinction. Early on in *The Imperative* he raises the issue of anthropocentrism, wondering whether our responsibility for life is a direct moral duty, or rather just an indirect requirement for the survival of human life (*IR*: 6–8). Several chapters later he answers his own question by stating that 'we can subsume both duties [to future generations and non-human life] as one under the heading "responsibility toward man" without falling into a narrow anthropocentric view' (*IR*: 136). We shall examine how Jonas comes to this conclusion.

In the pivotal section regarding non-human life's moral significance Jonas makes two arguments. Firstly, he offers the common-sense observation that 'care for the future of all nature on this planet [is] a necessary condition of man's own' (*IR*: 136). Of course, if this were the only reason for being concerned about the existence of a functioning biosphere, then his ethic would undoubtedly be anthropocentric, and philosophically indefensible given his identification of each living being as an end-in-itself. Jonas's second argument, however, moves beyond mere utility by way of a thought experiment: he asks that we consider a future wherein humanity had replaced non-human life with an artificial environment. The prospect of such a future focuses the mind on the possibility that 'the plenitude of life, evolved in aeons of creative toil and now delivered into our hands, has a claim to our care in its own right' (*IR*: 136). He justifies this as follows:

> [E]ven if the prerogative of man were still insisted upon as an absolute, it would now have to include a duty toward nature as both a condition of his own survival and *an integral component of his unstunted being*. We have intimated that one may go further and say that the common destiny of man and nature, newly discovered in the common danger, makes us rediscover *nature's own dignity and commands us to care for her integrity* over and above the utilitarian aspect. (*IR*: 137; emphasis added)

Jonas believes the two reasons emphasized justify the claim that his ethic moves beyond a narrow anthropocentrism, so we shall take them one at a time.

The first is somewhat ambiguous in that it is unclear whether having a duty as such (i.e. being a moral agent) is the integral component of a full human life, or if it is that the duty is *towards nature* that is essential. From the context it initially seems as though the latter is correct. This would, however, have the repugnant implication that the ecological crisis alone has allowed us to fully realize ourselves, since, for Jonas, prior to acquiring power over biosphere we could not have been responsible for it. For this reason I interpret Jonas's first reason as follows. Humanity's relation to the earth *is* essential to our being, but not because we now happen to be responsible for it. Rather, the earth is our 'worldly home in the most sublime sense of the word' because it is where humanity carves out its world, and realizes itself (*IR*: 137). As his great friend Hannah Arendt notes, '[t]he earth is the very quintessence of the human condition, [...] providing human beings with a habitat in which they can move and breathe without artifice' (1958: 2). According to this line of thinking, we *belong* to the natural world in the fullest sense of the word, and a humanity living within an entirely artificial environment 'would only result in the dehumanization of man, the atrophy of his essence' (*IR*: 136). A moral duty *to nature* is not, therefore, integral to being human, even if our existing on earth is.

This reading is also consistent with Jonas's wider thought. He says that the hypothetical artificial future dehumanizes us because it contradicts the reason for preserving humanity 'as sanctioned by the dignity of his essence': namely, *that* we are responsible (*IR*: 137). Destroying nature would mean destroying an object of our responsibility, and there would then be nothing to confront 'the arbitrariness of our might' (*IR*: 137). This would contravene the idea of Man *qua* responsible being, and *this* is the reason Jonas argues that nature has become for us an object of 'metaphysical' responsibility. As he puts it elsewhere, '[I]t is not so much a moral duty, it is really a metaphysical or ontological duty of ours to minimize our necessary, our on-the-whole inevitable [and] destructive impact on our environment' (*OS*: 29–30). The 'not so much' reiterates that there is a moral duty to life as such, but this is folded into responsibility for *our* future because it now 'comprises the rest under its obligation' (*IR*: 137). The argument thus escapes crude anthropocentrism by subsuming nature's survival under the idea of Man, hence Jonas's claim that the 'causes converge from the human angle' (*IR*: 136).

Jonas's second reason, which he thought fully evaded the charge of anthropocentrism, was that 'the common destiny of man and nature, newly discovered in the common danger, makes us rediscover nature's own dignity and commands us to care for her integrity over and above the utilitarian aspect'. The key notions to be extracted here are those of nature's 'dignity' and 'integrity'. Since these are somewhat vague terms, and Jonas does not define them precisely, his meaning must be inferred from the context. We shall see that doing so allows us to justify Jonas's grading of human beings above non-humans in the scheme of moral significance.

At various points in his writings Jonas refers to the dignity or integrity of living beings, species and nature as a whole, usually in criticism of modern materialism, which denies these entities such qualities. For instance, characterizing the Baconian world view, Jonas reformulates the famous line from Dostoevsky's *The Brothers Karamazov* '[w]ithout God [...] everything is permitted' (1992: 589), to read: 'If nature sanctions nothing, then it permits everything. Whatever man does to it, he does not violate an immanent integrity, to which it and all its works have lost title' (*PE*: 71). Jonas's use of integrity here refers, in the last clause of the latter sentence, to both nature as a whole and its 'works', which implies either species or individual organisms (or both). On the same theme, he says that scientific materialism has divested nature 'of any dignity of ends', and yet 'a silent plea for sparing its integrity seems to issue from the threatened plenitude of the living world' (*IR*: 8). By contrast, as part of his recasting of the Darwinian revolution as a blow to materialism (as recounted in Chapter 2), Jonas says the re-established connection between humans and nature meant 'some dignity had been restored to the realm of life as a whole' (*PL*: 57). Finally, there are references throughout his work to the dignity (*IR*: 46, 99, 197–8) and integrity (*IR*: 21, 34, 202) of human beings, in particular to the idea of Man.

We shall try to shed some light on Jonas's invocations of dignity and integrity, taking the latter first. Jan Vorstenbosch nicely captures the general sense of integrity, defining it as the '"wholeness", "intactness", and "unharmed or undamaged" state of something, presumably a living being' (1993: 110). Vorstenbosch's application is unnecessarily limited, however, since we can also speak of the integrity of collectives, such as a family or parliament, and even non-living beings such as a building or an artwork. What these beings have in common is that there is a particular way in which they ought to be. They have, in other words, a *telos*, either immanent or transcendent. Given that Jonas refers to 'immanent integrity' it seems he has the former concept in mind, and his application of it to both living beings and nature as a whole allows us to employ a distinction of Michael Hauskeller's and say that Jonas is here invoking 'biological integrity' and 'ecological integrity', respectively (2007: 37).

Now, the problem we encountered in Chapter 4 was that while it makes sense to speak of the immanent *telos* of an organism, one cannot say the same of the biosphere, which is not a self-organizing being but rather a *system* of beings whose actions affect one another, and only more or less maintain equilibrium. For this reason it seems that one may only speak of ecological integrity in the same sense as familial integrity, that is, a standard that is ascribed to the unit.

Immanent integrity therefore appears to apply only to individual living beings insofar as they are 'whole' and 'intact' in accordance with their *teloi* – and not, as Jonas would have it, the biosphere also. Which is then true of the integrity of the *idea* of Man – that is, the human being as responsible – invoked by Jonas? If there is no immanent integrity without a biological *telos*, it appears that the idea of Man, *qua* idea, also cannot be said to possess this status. It would presumably then follow that the integrity of the idea of Man is ascribed, somewhat like the moral integrity of a parliament or court of law. On this basis, at least, there appears to be no reason for prioritizing humanity in our moral considerations, since all living beings possess immanent integrity while the idea of Man does not. So while the appeal to integrity duly absolves Jonas of the charge of anthropocentrism, it actually undercuts his claim for humanity's 'superior right', which would require a strong qualitative distinction (*IR*: 137). This is the function of his use of dignity.

As mentioned, Jonas refers to both 'human dignity' and the 'dignity of ends' as such. Hauskeller again draws a useful distinction between 'personal dignity' and 'non-personal dignity' that neatly corresponds to the notions invoked by Jonas (2007: 63). We shall unpack these in turn.

Historically speaking, the two conceptions of dignity emerged from different traditions that over time became conflated under a single term. Personal dignity comes from the notion of *dignitas*, which human beings alone are supposed to enjoy by virtue of being made in the image of God: the *imago Dei*. The notion then underwent a secular revision in the thought of Kant, who says that 'morality is the condition under which alone a rational being can be an end in himself [...]. Hence morality and humanity, insofar as it is capable of morality, alone have dignity' (1993: 40–1). Jonas approvingly invokes Kant's conception of personal dignity (*IR*: 89), and not dissimilarly points to the capacity for responsibility as the criterion for being 'a member of the moral order' (*IR*: 99). Humanity's record in realizing its responsibilities is of course mixed; hence Jonas concludes that the 'dignity of man per se can only be spoken of as potential, or it is the speech of unpardonable vanity' (*IR*: 99). For Jonas, then, personal dignity, emerging from the tradition of *dignitas*, belongs to human beings alone insofar as they fulfil their capacity for moral responsibility.

The second tradition that constitutes our contemporary understanding of dignity is *bonitas*. *Bonitas*, according to Hauskeller, derives from the notion that 'everything created by God is good. [...] It emphasizes the community: the common needs, the vulnerability and mortality of everything that is alive' (2007: 63). In drawing on notions of need, vulnerability and mortality as shared properties of living beings, *bonitas* chimes with Jonas's invocation of the 'dignity of ends' belonging to living beings as such in their immanent-teleological constitution. In effect, then, dignity understood as *bonitas* appears to coincide with the idea of the organism as an end-in-itself, which similarly followed from the being of the organism.

By comparing these two traditions we can therefore make sense of Jonas's invocation of life's 'dignity of ends' as non-personal *bonitas*, and his references to human dignity as referring to our unique personal dignity, or *dignitas*. At face value this would indeed seem to give humans a qualitatively higher status than non-human life, insofar as they alone possess personal *and* non-personal dignity, whereas non-human life has only the latter to take into account in our moral considerations. Although Jonas does not sketch out the claim in detail, it does seem as though he holds that the 'higher', more existentially free the animal, the greater its dignity and capacity to be wronged (*OS*: 24–5). To be sure, this position has an intuitive plausibility, but there is a complication: we noted that Jonas drew on Kant's understanding of human *dignitas*. For Kant, this status was attributed on the basis of the person being 'an end in himself' courtesy of the capacity for moral agency (1993: 40). But Jonas then appears to attribute *bonitas* to the organism on the basis of *it* being an end-in-itself courtesy of its immanent teleology. If correct, this means that there can be no ranking of humanity as higher than non-human life: although humans alone have the status of personal dignity (*dignitas*), since this relies upon the human's status as an end-in-itself this would not make that form of dignity qualitatively different from non-personal dignity (*bonitas*), which is, according to Jonas, underpinned by the same.

It appears that several paths coincide in Jonas's thought. *Bonitas*, integrity and being an end-in-oneself all lead back to the possession of an immanent *telos* that characterizes life, while *dignitas* also seems to refer to being an end-in-oneself, despite having a separate ground in our moral being. As such, it seems that Jonas's claim that humanity's moral being 'surpasses all that has gone before with something that generically transcends it' is only half-correct: although responsibility corresponds to a uniquely human personal dignity – and in that sense is distinguished from the dignity of all other life – this difference only denotes another aspect of being an end-in-oneself.

This would leave us, as a result, with a resoundingly biocentric ethics, one that is egalitarian between different species and thus at odds with Jonas's original intentions. Of course, one might reasonably ask why we should *want* a principle that allows us to morally distinguish between human and non-human life. The answer is that a theory of living beings as ends-in-themselves cannot otherwise morally distinguish between a human being and a bonobo, even though the gulf that exists between the call of responsibility for these beings – both objective and subjective – is simply too great to ignore. The question is then whether, according to the interpretation of Jonas's philosophy provided, it is *possible* to justify an overriding obligation to humanity while respecting the dignity of all living beings.

One option is to follow Jonas's suggestion that the 'egoism of the species – each species – takes precedence [...] and the particular exercise of man's might *vis-à-vis* the rest of the living world is a natural right based on faculty alone' (*IR*: 137). At first this looks suspiciously like a 'might is right' argument. However, in a later lecture he supplements the idea that superior power legitimates its own use by arguing that there are nevertheless moral constraints on our conduct towards non-human life. Firstly, he suggests, we ought not to treat individual animals cruelly:

> And not out of respect, not out of awe, not out of admiration, but from simple decency. It is just indecent to relish in making a sentient being suffer merely for the sake of enjoying their suffering, or enjoying one's powers of destruction and inflicting pain. So this has little to do with the general question of how we should relate to the living world[;] it has more to do with the question of what kind of human beings we ought to be. (OS: 25)

This point appears, then, to be a virtue-based case against the characteristic of cruelty rather than one of responsibility for the integrity or dignity of non-human beings.

The latter only arises when Jonas discusses the *general* treatment of non-human life, in particular the methods of mechanized agriculture that deny animals a life in accordance with their immanent *teloi*, and thus their biological integrity.[12] So we can see that, at least in his later work, Jonas did not advocate the use of force against individual living beings with the sole condition that the biosphere remains intact. There are also questions of moral character, which Jonas claims follow from 'limiting the guilt which is ours by our dealings with the natural world' (OS: 24). One might ask, however, whether invoking such moral restrictions is really consistent with the general argument that 'our right to use all the rest of the living kingdom [...] simply derives from our power, because this is the only warrant in the whole realm of life which, as it were, entitles from the beginning a species to do what it does' (OS: 28). The latter claim holds for non-human species, to be sure, but *precisely because* they are not moral beings subject to duties and capable of virtue. For this reason we cannot accept it as a principle of human conduct.

The more fruitful alternative is to find an overarching principle allowing us to divide (formal) moral considerability from (relative) moral significance. And Jonas's philosophy *does* contain such a standard: the *nisus* of Being. Although we failed in Chapter 4 to rationally demonstrate the objective value of values, we were able to make some sense of the idea of Being's striving towards psychophysical freedom, moving from inanimate nature through to plant, animal and finally human life. To reiterate, on Jonas's schema this is not a true *telos* but rather the weaker notion that time enough and favourable circumstances, as on earth, will allow a tendency in Being to realize itself. As Jonas says, 'In organic life, nature has made its interest manifest and progressively satisfies it, at the rising cost of concomitant frustration and extinction, in the staggering variety of life's forms' (IR: 81). On this basis Being itself has a good of its own in that the fulfilment or frustration of its *nisus* allows for one to speak of a better or a worse for it. We can then say that it is better for Being that life came into existence than if it had not, and better again that vegetative and then sentient life followed. And of course, this *nisus* has led to humanity: the qualitatively significant feature of which is its moral being. This last point is crucial: the unique personal dignity of human beings, guaranteed by the capacity for morality, then becomes the crowning achievement of Being: 'This blindly self-enacting "yes" gains obligating force in the seeing freedom of man, [...] the supreme outcome of nature's purposive labor' (*IR*: 82). Jonas's philosophy of life thereby allows us to say that humans, insofar as they have the capacity for morality, are indeed of greater moral significance than non-human life according to a standard independent of any life form. It is thus human personal dignity, or the idea of Man as a responsible being, that rescues our interpretation of Jonas's ethics from a biocentric egalitarianism. If integrity and non-personal dignity (*bonitas*) indicate moral considerability, humanity's personal dignity – the capacity for morality – indicates something greater still: a cosmic moral significance.

Jonas's ethics thereby strikes an appealing balance between human and non-human interests. The demonstration of the integrity, non-personal dignity and status of being end-in-itself of all living beings fulfils the moral promise of his biocentric axiology: all these qualities make non-human life an object of responsibility in its own right, and thus his thought escapes the crude anthropocentrism which has underpinned so much of Western thought. At the same time, Jonas's ethics has a humanist dimension in affirming the transcendent value of the idea of Man, requiring the continued existence of humanity as its bearer, who for that reason occupies a privileged place in the hierarchy of moral significance. We may conclude, therefore, that Jonas's philosophy is axiologically biocentric, yet ethically weakly anthropocentric.

Where our interpretation of Jonas's position lacks intuitive appeal is in its remaining egalitarianism between the interests of animals and plants. The former cannot be more morally significant than the latter, since neither has the capacity for morality, and so neither obtains the status of personal dignity which allowed for the qualitative distinction between otherwise equivalent kinds of moral considerability. Jonas himself equivocates on this point. He argues that 'the gradings of world openness, capacities, modes of experience, modes of action, [and] modes of life' mean that 'certain instances of life [...] have higher value and greater goodness' (*OS*: 27). This may be so, according to the independent standard provided by Being's *nisus*, but we still cannot say that we owe more to a higher animal than a lower one, or a lower animal lower animal than a plant, since all possess only non-personal dignity. Jonas seemingly contradicts himself on precisely this issue, speaking of a 'grading of life as an underpinning for differentiating modes of behaviour towards them', yet also claiming that higher beings are *not* 'entitled to particular advantages' (*OS*: 27). We must conclude, on the basis of the arguments given, that only an egalitarianism between animals and plants stands to reason. This is regrettable, but we were at least able to uphold what Jonas correctly regards as the transcendent value of humanity amid the already value-rich landscape of Being.

What does this mean in practical terms? Once again, we must admit that Jonas's ethic provides us with a broad principle of action rather than a calculus for fine-grained decision making: such is the shortcoming of his theory, which would have to be supplemented with an adequate theory of rights and justice in order to be comprehensive. Nevertheless, *as* a broad principle it allows us to firmly conclude that non-human life deserves a future for its own sake, prior to the fact that its continued existence is necessary for our own. And for metaphysical reasons we must not replace the natural world with artifice: this would contravene the idea of Man as a responsible being, the continued existence of which is the object of our overriding obligations. For this very reason, however, we are relieved from the paralysing consequences of a biocentric egalitarianism. We are permitted to take non-human life to the necessary extent, even if in so doing, according to Jonas, 'we incur a certain amount of guilt': for 'life, as soon as it is manifold in form, is of necessity […] combined with conflict, and conflict [u]nto death' (*OS*: 23–4). Our task, in short, is to achieve a sustainable present so as to ensure a future for life on earth. This is accordingly the new task of politics in the age of technology, and the topic of Chapter 6.

6

The politics of nature

I The nature of politics

The concluding chapters of *The Imperative of Responsibility* are concerned with the political realization of our new planetary and intergenerational responsibilities. This shift towards practical considerations makes sense when we recall that Jonas's analysis of modern technology – though originating in a metaphysical revolution – addressed its socio-economic dimensions. The latter included not only market economics, which is subjected to further scrutiny here, but above all the gnostic eschaton of techno-utopianism. As such, Jonas's political theory is, at bottom, an attempt to confront this dimension of modern gnosticism and overcome it with a politics grounded in his ethic of responsibility.

As discussed in Chapter 5, Jonas's new ethic is intended to compensate for the fact that contemporary society's technological reach has spatially and temporally outstripped the scope of our historically given norms. This attempt includes, almost as a matter of course, reflection on the way in which his ethical principle might be put into practice. For '[i]f the realm of making has invaded the space of essential action, then morality must invade the realm of making, from which it has formerly stayed aloof, and must do so in the form of public policy. […] In fact, the changed nature of human action changes the very nature of politics' (*IR*: 9). Jonas's claim that morality has 'formerly stayed aloof' from production at first appears strange: the two are strongly connected in mixed and planned economies, and even in the United States, at the time of its foundation, economics was widely held to be subordinate to civic virtue (Sandel 1996: 123–200). We can, however, make sense of Jonas's point by levelling it at contemporary liberal democracies, which indeed allow the market a great degree of freedom.

Jonas's ethics, by contrast, generates a concrete obligation around which to organize action when transposed into the political domain. His attempt to do so results in two key insights: first, he devises an early formulation of the precautionary principle, and second, he recasts the state's duties along ecological lines. As such, Jonas politically prioritizes ecology over biotechnology, even though both pose a threat to the idea of Man. The reason is that the latter is a comparably specific and localized concern, whereas the former constitutes a global 'storm [...] that we, its unintentional creators, have the planetary duty of trying to avert' (*MM*: 51). It follows that although politics, like ethics, previously pertained only to the affairs of humans in states – the *polis*, or city – it must now reach beyond and account for the very earth that is the condition of possibility for such polities. Jonas writes that 'the boundary between "city" and "nature" has been obliterated', and, as such, '[i]ssues never [before] legislated come into the purview of the laws which the *total city* must give itself so that there will be a world for the generations of man to come' (*IR*: 10; emphasis added).

However astute Jonas's diagnosis will prove to be, his aforementioned solutions are controversial. The better received of the two is his formulation of the precautionary principle, which is now a generally accepted mechanism of environmental policy, at least in Europe. By contrast, Jonas's statecraft was and still is subject to a significant degree of opprobrium. His vision has been variously characterized as 'antihumanist' (Ferry 1995: 81), 'paternalistic' (Bernstein 1995: 17) and even an 'eco-tyranny' (Furnari 2006: 152). It cannot be denied that these criticisms have, at face value, some justification. However, on an alternative reading of Jonas's work all are the result of a basic interpretative error: according to Nathan Dinneen (2014; 2017), Jonas's flirtation with authoritarianism is merely a heuristic device, and not at all to be taken literally.

In this chapter we shall examine Jonas's political theory and come to a settled position on this debate, using a range of sources rather than just *The Imperative of Responsibility*. Doing so also reveals moments in Jonas's last works that point towards a different theory of citizenship and the state than that expounded in *The Imperative*. Although these suggestions and asides go undeveloped, a degree of extrapolation shows that Jonas was moving towards a theory of civic republicanism.[1] But we must first of all return to Jonas's philosophical anthropology and the place of politics therein.

Two political themes recur throughout Jonas's work: freedom and the *polis*. Jonas's interest in the latter is perhaps not surprising given his close friendships with Hannah Arendt and Leo Strauss, even if he rarely discusses their work.[2] The former theme, meanwhile, runs like a thread throughout Jonas's oeuvre: it is the focus of an early publication on St Augustine (*APF*), is covered in depth in the lecture course *Problems of Freedom* and of course, it characterizes his entire philosophy of life, as indicated by *Organism and Freedom*. The last is both Jonas's richest contribution to understanding freedom and the analytically necessary starting point, since the phenomenon is constitutive of organismic being as such.

We saw in Chapter 2 that Jonas identifies freedom as a feature of metabolism, which is the defining property of life. In reconstituting itself the organism obtains a formal independence from substance, although the ceaseless nature of this process and its reliance on material satisfaction mean that this freedom is far from unconstrained: it is instead a paradoxically needful freedom. If freedom is therefore formally co-existent with life, in tracking the various teleological means of satisfying metabolic needs it also admits of degrees. We recall that the freedom of the plant is apparently minimal, seemingly restricted to growth, photosynthesis and some motility (although we noted that the emerging science of plant 'intelligence' and communication may yet complicate this analysis). Animals, by contrast, possessing sentience, emotion and locomotion attain a progressively greater degree of freedom from the strictures of their environments, building a rich world. And finally, humans possess the greatest freedom of all life forms in their symbolically mediated existence, making images, speaking, thinking abstractly and radically reshaping their surroundings.

To clarify Jonas's usage of freedom here we may refer to its 'positive' and 'negative' variants, which he mentions only in passing (PF: 267). The classic distinction, generally attributed to Isiah Berlin, is as follows: positive freedom is the ability to act wilfully, while negative freedom is the absence of external interference. A third form of freedom, which Berlin does not consider, belongs to the civic republican tradition and conceives of freedom as non-domination. The relevance of this alternative will become apparent later; for now, the positive–negative distinction is sufficient to elucidate the freedom belonging to organismic being. Taking both together, we might say that the degree of an organism's freedom is the extent to which it can set itself ends and carry these out without external interference. Hence a rabbit is far freer than a weed in the *positive* sense, but a fox is *negatively* far freer than the rabbit on which it preys, despite the two mammals sharing many of the same capacities and so being roughly equally free in the positive sense. Problematically, however, Jonas's usage of freedom to describe life stresses the positive dimension alone. For him, organismic freedom is determined by life's process of individuation and the different modes of mediation to which it progressively gives rise. The importance of this oversight cannot be overstated: Jonas's proclivity to conceive of life's freedom solely in terms of the capacity to act, and not *also* in terms of non-interference or non-domination, has fateful consequences for his political theory, as we shall see.

Positive freedom is sufficient, however, to describe the human capacities that give rise to politics itself as a sphere of existence. In his philosophical anthropology Jonas stressed our creativity, demonstrated by the invention of fire, and our capacities for reason and metaphysics courtesy of our openness to the *eidos*. Together these form the conditions of possibility for politics, the latter being 'a new dimension of existence closed to animals other than man' (*PF*: 260). At the same time, should politics succeed in providing an ordered life it increases the opportunities to exercise our symbolically grounded capacities: as Jonas says, it 'affords [the] power to act otherwise than by dictate of sheer necessity', and is therefore the 'real phenomenon of freedom' (*PF*: 267–8). Although Jonas holds up the Greek *polis* and Roman *res publica* as the paradigmatic examples of political units, the concept is meant to far exceed the institutions and cultures of antiquity. Politics is for him a philosophical anthropological notion, akin to Aristotle's definition of humanity as 'by nature a political animal', but requiring only a symbolically organized life beyond the family unit (1984h: 1253a).

To reiterate, Jonas's understanding of politics here is as a *capacity* – a manifestation of our existential freedom – which does not logically entail any particular political *content*, nor, we might note, does it prohibit any. But there is nevertheless a worry that his biological and anthropological conception of freedom as purely positive, and not also negative, leads Jonas to endorse a positive theory of freedom when considering the state and citizenry. We shall return to this point later on, at present merely noting that the ontological ground of politics is humanity's symbolic existence, which is itself the highest manifestation of life's movement towards freedom. We shall now look at how Jonas substantiates his formal analysis of politics with his ethic of responsibility, leading first of all to his influential formulation of the precautionary principle.

II New rules for collective action

As discussed in Chapter 5, Jonas's categorical imperative holds that we must safeguard the existence of human beings on earth in accordance with the idea of Man. To bring this ethic to bear on the realm of public policy, Jonas calls for a 'science of hypothetical prediction, a "comparative futurology"', to underpin a 'heuristics of fear' which might guide our actions (*IR*: 26, x). In introducing such a dubious sounding concept, it should be noted straight away that the heuristic of fear is supposed to respond to scientific, rather than arbitrary, predictions. Jonas has in mind descriptive analyses of 'presently recognizable trends in the technologic-industrial process' allowing for the forecasting of 'certain, probable, or possible outcome[s]' (*IR*: 26, 30).

We might then ask who would make such assessments. Jonas suggests that 'the biologist, the agronomist, the chemist, the geologist, the meteorologist, [...] the economist and engineer' could pool their knowledge to form a 'global environmental science' (*IR*: 189). Their predictions, he says, would have to be of 'a still higher degree' of rigour than that 'which suffices for the short-range prediction intrinsic to each work of technology by itself', because such a forecast 'is on principle inadequate for the long-range prediction' (*IR*: 28–9). What the criteria are for this rigour, and who decides upon them, is not addressed, this presumably being a matter for the scientific community. But the end result might, perhaps, look like a more ambitious version of the UN's IPCC.

Of course, policy does not rest on scientific knowledge alone, but also concerns what is and what is not desirable. As such, the heuristic of fear would also utilize narrative to bring hypothetical situations to life. In this connection Jonas mentions Aldous Huxley's novel *Brave New World*: a 'well informed thought-experiment' pertaining to biotechnology and which to this day has not lost its power to repulse (*IR*: 30). The particular value of such works – provided they are scientifically and technologically plausible – is in 'developing an attitude open to the stirrings of fear in the face of merely conjectural and distant forecasts – a new kind of *éducation sentimentale*' (*IR*: 28). The moral legitimacy of doing so follows from Jonas's theory of responsibility, which sought to not only be rationally demonstrable but also to account for the *feeling* of responsibility for the vulnerable good. Hence we are not to be guided by a 'pathological [...] but rather a spiritual sort of fear' for what might be lost through human (in)action: both the vulnerable good itself and the idea of Man in failing to fulfil our obligation to care for that good (*IR*: 28).

However, even if one accepts Jonas's elision of the descriptive and the normative, one might ask why he focuses on fear for the future and not, say, hope. His reason for doing so is reminiscent of the moral psychology advocated by Hobbes, 'who also, instead of love for the *summum bonum*, made fear of a *summum malum* [...] the starting point for morality' (*IR*: 28). Hobbes' application of this principle pertains to humans in the state of nature, who, being in 'continuall feare, and danger of violent death', are driven to peaceful cooperation and the establishment of civil society ((1651) 1914: 65–6). For Hobbes, in other words, it is each individual's fear of what might be lost – their life – which alerts them to what is most valuable: security.

According to Jonas this is in fact an example of the pathological fear that is unsuitable as a basis for collective decision making, as individual security, while of great value, is not an absolute never to be risked. There are, after all, things worth dying for. The general structure of the principle, however, remains sound: '[T]he perception of the *malum* is infinitely easier to us than the perception of the *bonum* [...] an evil forces itself upon us by its mere presence, whereas the beneficial can be present unobtrusively and remain unperceived, unless we reflect on it' (IR: 27). In the case of imagined but realistic future scenarios, we are led to that object within which we must not risk: the existence and essence of humanity. Having perceived the threat, brought to our attention by the faculty of fear and applied to an object whose worth we can rationally account for, we may then act so as to not bring that harm about. Thus '[t]he prophecy of doom is made to avert its coming, and it would be the height of injustice later to deride the "alarmists" because "it did not turn out so bad after all". To have been wrong may be their merit' (IR: 120).

Jonas admits that the 'uncertainty of prognostications' poses a problem for his heuristic (IR: 28). In short, when applied to hypothetical situations it only prohibits actions of which we can be more certain than not of the consequences. For example, it is highly likely that a nuclear war would make truly humane life impossible – and perhaps even extinguish human life itself – and so it clearly falls foul of his method. However, there are technologies that present us with great uncertainty as to their possible effects. For example, the development of general artificial intelligence could lead to sustainable material security for all, or, at the other extreme, to the effective obsolescence of humanity. The problem is that we cannot predict with much certainty which is correct: the benefits and harms are at this stage too remote and conjectural. How then are we to know which assessment we should side with? To deal with this impasse Jonas offers an early formulation of a new rule by which to guide collective high-stakes actions in the face of uncertainty: the precautionary principle.

In the event that a particular action or technology poses the threat of catastrophe – however uncertain the threat, and benign the other possible outcomes – Jonas recommends that we give precedence to 'the bad over the good prognosis' (IR: 31). The reason harks back to his categorical imperative: any action in which either the existence or essence of humanity is at stake is ruled out in advance. Now, one common reductio ad absurdum of the precautionary principle is that it rules out *any* action that poses a great and irreversible threat to life: travelling by aeroplane, say, or even crossing the road in busy traffic. Certainly, either activity risks an irreversible loss of life, but to say that precautionary logic rules them out is a wilful misunderstanding of the principle, or at least of Jonas's version of it. Individuals and groups have justifiably risked their lives throughout history, and there is no reason why contemporary technology should change this. What contemporary civilization *does* endanger, in extreme cases, is the existence of humanity as such, and it is *this* that Jonas's precautionary principle is designed to counter. Hence Jonas states – even coming close to using the phrase 'precautionary principle' – that 'we must bow to the command to allow, in matters of such capital eventualities, more weight to threat than to promise and to avoid apocalyptic prospects even at the price of thereby perhaps missing eschatological fulfilments. It is the *command of caution*' (*IR*: 32; emphasis added).³

If the purpose of such precaution is the avoidance of infinite loss, its price is the renunciation of finite gains. In this respect his theory is reminiscent of Pascal's wager, albeit with one key difference. Pascal's argument famously holds that it is better to live a life according to God than not, as the cost of doing so and being wrong is minimal compared to wrongly doubting God's existence and being subsequently condemned to eternal damnation. We might say that Jonas's theory likewise bids us to avoid catastrophe in the future by making a comparably minimal sacrifice now. The difference is that unlike in Pascal's version, where heavenly bliss is one possibility, there is no future utopia to be had, only the preservation of the status quo. However, if Jonas is correct, this is no reason for despair: the avoidance of the infinite loss in the future presupposes that we possess something of absolute worth *now*, namely, human life on earth. For all the overwhelming (but necessary) negativity of his formulation, Jonas's precautionary principle is therefore also motivated by an affirmative rationale: '[A]n emergency ethics of the endangered future must translate into collective action the "Yes to Being" demanded of man by the totality of things' (*IR*: 140). That is, the *nisus* serving as the ultimate proof of the value of Being is to find its political expression in the precautionary principle.

Perhaps the only English-language commentator to recognize Jonas's influence on the precautionary principle is Kerry Whiteside, and he raises key questions about Jonas's formulation of it.⁴ Above all, Whiteside observes that the largely Kantian basis of Jonas's conception – the new categorical imperative – means that although we are unconditionally barred from taking certain actions, these actions are in practice relatively few. In other words, what we gain in clarity we lose in breadth of application, as 'it is made in such a way that it almost never applies' (2006: 108). There is something to this objection: as noted, nuclear warfare is one example that falls foul of Jonas's formulation, but we hardly need a new principle of action to know that we should not start one. By contrast, other technologies that are often seen as paramount cases for a precautionary approach do not fall under the remit of Jonas's version of the principle. Take agricultural genetic engineering: opponents sometimes claim that a proliferation of genetically modified crops and livestock could be damaging for biodiversity and the global food supply. Crucially, such effects are possibly irreversible as the intervention is designed to be hereditary. But no opponent of GMOs, however strident, believes that they risk the end of human life on earth or humanity as a responsible being, as Jonas's principle demands for a precautionary veto.

We could pick a number of other technologies – nuclear power, geoengineering, synthetic biology – the widespread adoption of which could have disastrous ecological effects and yet not come under the remit of Jonasian precaution. It is presumably for this reason that where the precautionary principle has been adopted in law it is in a weaker but more broadly applicable form. Take, for instance, Principle 15 of the UN's Rio Declaration on Environment and Development: 'In order to protect the environment, the precautionary approach shall be widely applied by States according to their capabilities. Where there are threats of serious or irreversible damage, lack of full scientific certainty shall not be used as a reason for postponing cost-effective measures to prevent environmental degradation' (United Nations 1992). The advantage this strictly consequentialist formulation has over Jonas's is that it applies to threats of serious or irreversible harm to the *environment*, and not just when the existence or essence of humanity is at stake. However, the stipulation that the principle be 'widely applied' by states – which is already vague – is further weakened by the caveat 'according to their capabilities', just as the remit of precaution is limited only to uncertainty not being 'used as a reason for postponing cost-effective measures'. Clearly this formulation also has serious flaws, albeit of a different sort to Jonas's.

A slightly more robust consequentialist version of the precautionary principle can be found in Article 11b of the earlier UN World Charter for Nature, which states: 'Activities which are likely to pose a significant risk to nature shall be preceded by an exhaustive examination; their proponents shall demonstrate that expected benefits outweigh potential damage to nature, and where potential adverse effects are not fully understood, the activities should not proceed' (United Nations 1982). The principle as it is expressed here still poses problems. The first clause refers only to those technologies *likely* to pose a significant risk to nature, rather than posing a significant risk *simpliciter*. Then the second clause seeks to weigh 'expected benefits' against 'potential damage to nature', meaning that only the latter phrase explicitly refers to the environment or non-human life whereas the former could be taken to encompass just human interests. As such, any activity that overwhelmingly benefited humans, even if it were at the expense of the environment, might not be ruled out. Moreover, the final clause – 'the activities should not proceed' – replaces the mandatory 'shall' from the first clause with the normative 'should', which is not synonymous. Nevertheless, the value of such a formulation is that at least some potentially harmful technologies, such as those mentioned earlier, are brought under its remit.

This brief comparison of varieties of the precautionary principle shows that Jonas's strict but narrow formulation is not the only version available. It is, however, the only one that follows from his premise of the new categorical imperative. Does this pose a problem for Jonas? Surely not. Firstly, it is simply not the case that 'critics are right to dismiss versions [of the precautionary principle] that follow a logic analogous to Pascal's wager, because that argument requires the assumption of an infinite catastrophe, which is seldom, if ever, the case in environmental decisions' (Johnson 2012: 9). We might be justified in dismissing a principle if it *never* applied in real-world scenarios, but why should we do so if it *does* apply in some cases, albeit rarely? Secondly, remember the scope of Jonas's imperative of responsibility: to supplement, rather than replace, traditional morality. Conventional consequentialist reasoning can justify a broader but weaker version of the precautionary principle, whereas Jonas's version categorically tells us what we must not put at stake; the two are complementary, not mutually exclusive. And finally, recall that the level at which Jonas envisioned the application of his principle was not just that of individual technologies, but also the general trajectory of contemporary civilization, which is indeed bound for catastrophe. We turn now to how this is to be institutionally enacted.

III Farewell to utopia?

The main political problem facing Jonas's heuristic of fear and precautionary principle is the reluctance of political parties to adopt such radical policies, however reasonable they may be. How then, according to Jonas, are the recommendations derived from the heuristic of fear to be transmitted into concrete action? Which political and economic systems are most capable of doing so?

Given the context of the Cold War in which he was writing, Jonas develops an answer by examining the relative virtues of the two systems available at that time: capitalism and communism. His approach is reminiscent of Aristotle's *Politics*, which proceeds by comparing the political systems of ancient Greece. The difference is that Jonas's investigation is not guided by 'the question as to which of the various political ideologies and programs is inherently best for human beings', which he dubiously claims is now 'not significant' (*M*: 202, 210). Instead he asks only 'which offers the greatest likelihood of meeting successfully the completely new challenge confronting human society: how we can live with nature – or how nature can survive together with us' (*M*: 210). From these remarks it is easy to see how Jonas arrives at such a controversial vision of the state, a vision that, it must be said, draws inspiration from some of the most questionable political theorists in the Western canon: not only the aforementioned Hobbes but also Plato, Machiavelli and Lenin. In Jonas's defence, however, he firstly develops an astute critique of the techno-utopianism built in to modern civilization, both capitalist and communist.

As discussed, part of technology's interaction with socio-economic life is at the level of ideology. More specifically, Jonas identifies modern gnosticism's eschatological dimension in the belief that utopia lies in the scientific and technological transformation of nature. In addition to this explicit articulation, Jonas diagnoses an implicit utopianism in the productionism of modern civilization. As discussed in Chapter 1, through competition, innovation and ever-greater consumption, creating new 'needs' where previously there were few, we are supposed to arrive at material satisfaction. We have for decades now been aware that such a vision sows the seeds of ecological ruin, and yet contemporary civilization carries on as though there were no limits to such activity. Thus what was in Bacon's time merely 'Promethean arrogance' has become a wilfully blind utopianism (*IR*: 143).

Although a productionist tendency belongs to both of the political systems he examines – the capitalist West and the then-communist East – he regarded it as stronger in the former, indeed, as almost synonymous with capitalism, which is characterized (or perhaps caricatured) as 'the unrestrained use of the world's resources, of the environment, of nature, impelled by the pocket-motive and competition' (*CR*: 217). As such, Jonas is sceptical on *a priori* grounds that it could rise to the challenge, demanded by his imperative, of averting the course of disaster.

What of Soviet communism, the only readily available alternative at the time?[5] Jonas notes that the moral force of Marxist-Leninism is that it 'proposes to bring the fruits of the Baconian revolution under the rule of the best interests of man' (*IR*: 143). Of course, this still poses a fundamental problem since the Baconian revolution is precisely the source of our ecological predicament. The fact that the techno-utopian drive is here inspired by a sense of distributive justice is commendable, but little help regarding the question of ecological limits. Jonas takes Ernst Bloch as the 'foremost prophet' of this kind of Marxism, since the theme of utopia is so explicit in his work and acts as a foil for Jonas's politics of responsibility (*IR*: 188).[6] Bloch's key work, *The Principle of Hope* (1986), envisions a society freed from the necessity of human labour through the rational application of technology and equitable distribution of goods. Jonas criticizes Bloch's vision partly on principle, regarding the transfer of meaningful work to automated industry as unbefitting of humanity's creative capacities (*IR*: 197–201). But the relevant objection here is the practical one of consumptive limits: if the earth cannot withstand continued capitalist production, then neither can it contain an 'onslaught on resources' in the name of utopian communism (*IR*: 187). Hence the 'dawning truth of ecology puts a hitherto unknown damper on progressivist faith, socialist no less than capitalist' (*IR*: 189).[7]

The virtues of the Soviet communist system, if any are to be found, will then be in its ability to constrain rather than promote the Baconian ideal. Jonas argues, again on *a priori* grounds, that Marxist-Leninism holds the 'promise of a greater *rationality* in the management of the Baconian heritage' than capitalism does (*IR*: 145). Empirically, of course, this belief has to contend with a woeful record of bureaucratic inefficiency and the fact that 'Marxism in one country' finds itself at odds with actively – or at least ideologically – hostile foreign powers, and is almost necessarily driven to raise production in response. Even if we suppose that there were a communist world state (hardly an imminent possibility, even in the 1970s), Jonas notes that centralization of the sort associated with a command economy would require efficient infrastructure, communications and bureaucracy, and this alone could be sufficient impetus for technological development and economic growth (*TPT*: 36).[8] To all this we must add that following the collapse of the Soviet Union its unprecedented degree of environmental mismanagement became clear – this being, for Jonas, 'one of the great disappointments' of the Soviet experiment (*CBE*: 29).[9] Prior to the emergence of this latter evidence, however, he thought it better able to act in accordance with ecological limits than a capitalist economy.

Jonas's central political claim in *The Imperative of Responsibility*, then, is that we should opt for a Marxist-Leninism shorn of its utopian productionism. And with these means must go their envisioned end: a communist society composed of 'true', that is, emancipated, humans, this being the theory's 'noblest and hence most dangerous temptation' (*IR*: 156). Why dangerous? Precisely because it bids us to forego what is of ultimate value – the existence and essence of humanity – for an imaginary perfected form of that being. Such eschatological promise justifies virtually all means, not least the Baconianism that unwittingly puts that very end at risk. Jonas certainly does not reject the egalitarianism of Marxist-Leninism, which he claims ought to be preserved (*IR*: 144). But its productionist aspect contravenes the heuristic of fear by imperilling that which must not be imperilled: genuine humanity, which 'is always already there' in having the capacity for morality (*IR*: 200). Protecting *this* being is the object of a politics of responsibility, necessitating the abandonment of radical hope for material emancipation, and leading Jonas to advocate a 'post-Marxist', or anti-utopian, form of Marxist-Leninism (*IR*: 127).

Jonas's preference for an ecologically minded command economy over the free market is acceptable in itself. Although he takes the Soviet Union as his paradigm case, there is no *necessary* connection between production for need and totalitarianism, and so nothing said in praise of an austere form of Marxist–Leninist economics should entail approval for its political model. But unfortunately this is not a distinction Jonas upholds. As Walter Weisskopf notes, by framing his discussion in terms of the central Cold War belligerents, Jonas pairs capitalism with democracy and socialism with dictatorship (2014: 32). And, since Jonas is concerned with which system is in principle best for the relationship between humanity and the natural world, rather than for human beings *alone*, Marxist-Leninism appears to offer better prospects. But there is another reason why Jonas arrives at this unhappy conclusion, which is that he takes his theory of responsibility to entail, almost by logical extension, a paternalistic form of government. We shall see if this is really the case, and how it feeds into his acquiescence to an authoritarian politics.

Jonas draws a parallel between private and public responsibility, between the responsibility of the parent for the infant and that of the statesman for their citizens.[10] The basis of his comparison is as follows: firstly, both have other humans as their object, in accordance with his imperative. Secondly, both pertain to the 'totality', 'continuity' and 'future' of those beings (*IR*: 98). The quality of totality refers to 'all aspects' of the object of responsibility, 'from naked existence to highest interests' (*IR*: 101). In the case of the infant this makes sense, but regarding citizens is surely too strong a claim. In defence of it Jonas cites the authority of Aristotle, who argued that the state 'came into being so that human life would be possible, and continues in being so that the good life is possible', from which Jonas concludes that this 'is also the object of the true statesman' (*IR*: 101). Clearly, however, the fact that the state exists to make the good life possible does not entail that it is concerned with every aspect of citizens' lives. Backtracking somewhat, Jonas then claims that any political executive who leads public opinion rather than follows it upholds something of the statesman's ideal. But once again, such leadership is clearly not equivalent to total parental responsibility. Perhaps the most we can say is that in securing the body politic from outside threats, ensuring law and order and providing access to education and the arts, the statesman has *some* – not total – responsibility for citizens' lives.

The second quality is continuity. As with totality, continuity follows from the vulnerability of the body politic: '[T]he insistent knowledge that the *res publica* too exists precariously' (*IR*: 104). For this reason it 'cannot allow itself a vacation or pause, for the life of the object continues without intermission' (*IR*: 105). Here Jonas's comparison appears sound, as in neither case is the responsibility periodic. Regarding the final quality, however – the future of the object of responsibility – it is again questionable. The parent's responsibility for the infant is continuous only up to the point of maturity, and in accordance with this immanent *telos* must gradually relinquish its claim to totality, as Jonas himself points out (*IR*: 108). Responsibility for the body politic, by contrast, is unending – though, as stated, never total – and passed on from one government to the next.[11]

The final similarity Jonas draws is that in both cases the responsibility is also to the *future possibility of there being responsibility*. As we saw with the parent, their responsibility is directed towards the idea of Man as a responsible agent embodied in the infant. Jonas argues that the statesman, too, in taking up a position of responsibility for the polity is duty-bound to ensure the continuation of statesmanship: that is, those who are *politically* responsible. Echoing his earlier categorical imperative Jonas writes: '[T]here follows a highly general, but by no means empty, *imperative* precisely for the statesman […] to do nothing that will prevent the further appearance of his like' (*IR*: 118). Here we are inclined to agree with Jonas, insofar as the finitude of the human condition requires any political leader to prepare, or at the very least not obstruct, others who will eventually take their place. To do otherwise would risk the future of the polity for which they are responsible.

If Jonas underestimates the extent to which the qualities of parental and political responsibility differ, he recognizes at least one significant difference: the self-chosen nature of the statesman's responsibility. Jonas notes that 'nobody is formally bound to run for public office' and take up the mantle of responsibility for the community pre-existing them (*IR*: 96). The infant, by contrast, having been brought into existence by the parent makes of them an irrevocable call of responsibility. Even so, Jonas goes on to blur the distinction by claiming that 'he who feels the calling for [leadership] in himself seeks the call and demands it as his right' (*IR*: 96–7).

Jonas's invocation of a political call again echoes Heidegger's discussion of the call of conscience in *Being and Time*. For Heidegger, as indicated in Chapter 5, the 'call does not say anything, does not give any information about events of the world, has nothing to tell. [...] "Nothing" is called *to* the self which is summoned, but it is *summoned* to itself, that is, to its ownmost potentiality-of-being' (2010a: 263). Precisely this formal understanding of conscience allowed Heidegger, Jonas claims, to align himself with the Nazis when he felt such a call. For in Heidegger's philosophy what matters is not '*for what* or *against what* one resolves oneself, but *that* one resolves oneself' (*HRR*: 201). In an excoriating passage Jonas says:

> Heidegger's being [...] is an occurrence of unveiling, a fate-laden happening upon thought: so was the Führer and the call of German destiny under him: an unveiling of something indeed, a call of being all right, fate-laden in every sense: neither then nor now did Heidegger's thought provide a norm by which to decide how to answer such calls – linguistically or otherwise: no norm except depth, resolution, and the sheer force of being that issues the call. [...] Heidegger's own answer is, to the shame of philosophy, on record and, I hope, not forgotten. (*PL*: 247)

According to Jonas's analysis of the political call of conscience, by contrast, '[t]he object of the responsibility is the *res publica*' and the good life of its citizens – *all* citizens – that it exists to promote (*IR*: 96). Thus his explicitly ethical interpretation of Heidegger's analysis of care finds its political counterpart in his explicitly ethical account of the call of conscience.

Returning to the issue at hand, we may conclude that Jonas was wrong to say that 'common traits' of parental and political responsibility 'make them blend into [...] the primordial phenomenon of responsibility' (*IR*: 98). The one arises in witnessing the vulnerable infant while the other – assuming it can be properly understood as a call – emanates from the collective of which one is a part. This distinction goes to the heart of their difference: responsibility for the infant rests on a fundamental asymmetry of power and vulnerability between the two parties (at least initially), whereas the statesman emerges from a group of equals to which they return in due course. It is thanks to this pre-existing equality that the statesman's responsibility lacks totality, and should instead be understood only as a temporary suspension of previous relations.

As discussed, however, Jonas does not fully acknowledge this and is happy to more or less draw an equivalence, even going so far as to argue that in the case of children both forms of responsibility coincide. Insofar as to rear a child is also to rear a citizen, the role played by education therein means that the state does both, and can even assume total responsibility for the infant in the case of parental neglect. This might not be a problematic observation in itself, but what *is* questionable is that Jonas does not say what an inappropriate degree of state intervention in child-rearing would be. Considering the communist argument for abolishing the family altogether, he suggests that 'this extreme case only magnifies what we assert about the responsibility of the statesmen in general and its affinity to that of parents' (*IR*: 103). There can be no doubt, therefore, that Jonas finds a paternalistic form of government at least palatable.

The combination of several factors – freedom understood only positively, tolerance of paternalism and the endorsement of a parsimonious Marxist–Leninist command economy – leads Jonas's political theory to a distasteful conclusion. Considering again the advantages of the Soviet system to reign in Baconian productionism, he counts 'total government power' among them – the only stipulations being that it must be 'well-intentioned [and] well-informed' (*IR*: 146–7). The reason given is that 'decisions from the top, which can be made without prior assent from below, meet with no resistance (except perhaps passive) in the social body and [...] are assured of implementation' (*IR*: 146). Such decisions could include, crucially, those of productive and consumptive austerity that run counter to the immediate self-interest of citizens, and would therefore 'be difficult to get adopted in the democratic process' (*IR*: 146). One example given, apparently sincerely, is China's one-child policy: '[A] shining example of what a communist regime can accomplish' (*IR*: 152). In his defence, Jonas regards this power as advantageous *only* if we can trust an authoritarian government to take the right course of action. He acknowledges that failure to use such power wisely risks far worse outcomes than capitalism is capable of, yet overlooks the fact that such failure is distinctly likely in an oppressive form of government.

If this were not concerning enough, Jonas asks how loyalty to such a government might be managed. As stated, the sorts of policies Jonas regards as necessary for the survival of human and non-human life on earth are unlikely to be popular with the general public, at least without immediate evidence before our eyes of the consequences of our productionist form of life. And if by 'immediate evidence' we mean that the tides must begin to rise in London and New York before we are spurred on to tackle the ecological crisis, then it will of course be far too late. Hence, for Jonas, the Soviet communist system once again appears advantageous, as 'only a maximum of politically imposed social discipline can ensure the subordination of present advantages to the long-term exigencies of the future' (*IR*: 142). The sort of social discipline Jonas has in mind is not that of violent repression, but rather the ideological cultivation of a public 'spirit of frugality' (*IR*: 147). In a Machiavellian vein, Jonas suggests that the tools of propaganda employed in the Soviet Union could be used to inspire not productive utopianism but the very reverse: 'enthusiasm for austerity' (*IR*: 148). And if the propagated truth of our ecological predicament fails to inspire such action, then he claims that government would be required to engage in a sort of political mythology, convincing citizens that the ascetic society *is* the good society. He speculates that '[p]erhaps this dangerous game of mass deception (Plato's "noble lie") is all that politics will eventually have to offer: to give effect to the principle of fear under the mask of the principle of hope' (*IR*: 149).

Jonas thereby arrives at eco-authoritarianism as the most plausible alternative to ruin. It cannot be stated forcefully enough that Jonas does not regard such governance as ordinarily desirable. On the contrary, when considering what constitutes the best state for human beings alone Jonas points to democratic government, civil liberties, nationalized industries and a welfare state – that is, European social democracy (*IR*: 174–5). Unfortunately, since he is comfortable with the notion of paternalistic governance, and concerned above all with how to rein in productionism and thereby ensure that there *be* a good future for human beings, democracy appears in that light as an acceptable, if regrettable, sacrifice.

Here Jonas betrays an uncomfortable affinity to Heidegger's post-war political statements. After Heidegger had ceased actively promoting the Nazi cause – never publicly renounced – he remained openly sceptical about the value of democracy. In his infamous *Der Spiegel* interview, Heidegger says: 'A decisive question for me today is how a political system can be assigned to today's technological age, and which political system would that be? I have no answer to this question. I am not convinced that it is democracy' (1990: 54). Many years later, also in an interview with *Der Spiegel*, Jonas comes to a strikingly similar conclusion: 'I too suspect that democracy in its present state, with its short-term orientation, is not a suitable form of government in the long run. And why should it be?' (*CBE*: 25). So we have here a great irony: in spite of his fierce criticisms of Heidegger's politics and their lack of ethical foundation, the political theory Jonas arrives at is, in one important respect, not so different from his teacher's later public position.

In this failure Jonas betrays the lingering influence of gnosticism's third facet. To see why, recall Voegelin's identification of how the gnostic principle is at work in modern politics. The 'immanentization of the eschaton' has, he says, two dimensions: the notion that there is a *telos* to history, discernible only by the few, and that this *telos* leads to a paradisiacal end-state. Although the goal of Jonas's politics is certainly not utopian, and to that extent clearly evades the ecologically ruinous gnosticism of Bacon and Descartes, in his analysis of statesmanship Jonas does conform to the tendency. The proper statesman, according to Jonas, is not only an exemplary individual but also one who *correctly recognizes the truth and governs in accordance with it*. And should that truth elude the *polloi*, led astray by their material wants, the statesman is obliged to govern against the express interests of the masses in the name of their true interests. The parallel here with the gnostic division between those who in life observed Holy Writ, and those who worshipped false idols, should be clear; indeed, Voegelin notes that this very aspect of gnostic movements is what leads them so easily into totalitarianism (1952: 132). It is ironic, then, that Jonas – whose work influenced Voegelin – should succumb to this aspect of gnosticism.

IV Rival interpretations of Jonas's politics

Can Jonas's political ecology be saved, and if so, how? As stated earlier, Nathan Dineen suggests one original – though ultimately unconvincing – way of doing so.[12] He stresses the role played by the heuristic of fear in Jonas's political thought, and argues that the subsequent analysis of authoritarianism can only be understood in light of it. He specifically points to a passage preceding the discussion of Marxist-Leninism's advantages over capitalism, where Jonas says: 'All this holds on the assumption made here that we live in an apocalyptic situation, that is, under the threat of a universal catastrophe if we let things take their present course' (*IR*: 140). For Dineen this reveals Jonas's true and much-misunderstood intention: that if we fail to prevent ecological collapse, then tyranny will force itself upon us by necessity, and it is therefore our duty to *envision this possible outcome precisely to avert it*. In this way Jonas can be understood not as endorsing authoritarianism, but in fact the very opposite: engaging in the kind of 'well-informed thought experiment' that his heuristic demanded so as to alert us to the likely terrible outcome of our present course of action. In Dineen's words: 'Jonas uses political dystopianism to counter the possibility of a political dystopia from coming into being' (2014: 18).

Dineen's interpretative twist finds textual support in a lecture (which he does not cite) where Jonas says the following:

> My dire prognosis [–] that not only our material standard of living but also our democratic freedoms would fall victim to the growing pressure of a worldwide

ecological crisis, until finally there would remain only some form of tyranny that would try to save the situation [–] has led to the accusation that I am defending dictatorship as a solution to our problems. I shall ignore what is a confusion between *warning* and *recommendation*. [...] This is, I want to emphasize, a worst-case scenario, and it is the foremost task of responsibility at this particular moment in world history to prevent it from happening. (*MM*: 111–12; emphasis added)

We may also point to a late interview in which Jonas suggests that '[i]n a lifeboat situation all rules cease to apply', and *therefore* 'we must prevent that lifeboat situation from coming about' (*IHJ*: 367). It appears, then, that Dineen is correct to say that Jonas's discussion of tyranny is an application of his heuristic of fear to the domain of political theory. On this reading the true meaning of *The Imperative* only emerges when considered as a whole, with the key argument of its second chapter – that 'the creatively imagined *malum*' can 'instil in us the fear whose guidance we need' (*IR*: 27) – explaining why a terrible political future is envisioned in its fifth.

The suspicion lingers, however, that Jonas's explanation just cited – given at a conference organized by the social-democratic Friedrich-Ebert-Stiftung – may simply be an *ex post facto* excuse in response to the heavy criticism he received. The main reason for thinking so is the absence of evidence in *The Imperative* that Jonas's advocacy of tyranny was not literal. The sole passage Dineen points to as evidence could easily be read as meaning that since we live in an 'apocalyptic situation' *right now* the political recommendations are meant accordingly. Then there is Jonas's admission that he does not 'stand aghast at the thought' of using a noble lie to lead the population into austerity (*IR*: 149). This does not sound like part of an elaborate thought experiment, but very much sincere, which brings us to a fundamental problem with Dineen's reading: it becomes unclear when we can take Jonas at face value and when we must assume he is speaking heuristically. For instance, the discussion of the statesman's responsibility – which as we have shown is clearly paternalistic, paving the way for authoritarianism – is presented by Jonas as following on principle from his theory of responsibility. At what point in this chain does the argument cease to be serious? For all these reasons it seems more likely that Jonas's arguments in *The Imperative* are indeed literal.

Taking together various remarks, his post-*Imperative* position on the necessity of authoritarianism appears to be as follows. Ecological collapse is one possible future but not yet a certainty: hence 'at this moment fatalism is a deadly sin' (*FWT*: 54). For one thing, preliminary ecological shocks – the initial tremors before the earthquake – may well spur us on to act before it is too late.

Harking back to the heuristic of fear, Jonas speculates that this might be enough to prevent the worst-case scenario of environmental turmoil leading to an eco-dictatorship: 'What I can imagine [...] readily is an outbreak of dire events leading to compromise among economic, political, and social power groups who would then reach an arrangement that is relatively acceptable in terms of people and planet' (*CBE*: 25). In the event that we should fail to take such action, however, he is willing to make the 'terrible concession' that 'tyranny would still be better than total ruin' (*MM*: 111–12). It would in those circumstances alone be acceptable for a government to 'employ such means, which we now abhor or at least deplore, in order to save their own existence' (*IHJ*: 366). Such a system is therefore now conceived of as a last resort, preferable only to an alternative of ecological collapse precipitating a new 'stage of primitivism', of 'mass poverty, mass death and mass murder, the loss of all treasures that spirit has produced' (*CBE*: 22).

Jonas's final caveat is that if an eco-dictatorship should prove necessary, he hopes that this will be a temporary measure only, holding freedom in trust until such a time as it might be allowed to flourish once more:

> We can make a terrible concession to the primacy of physical survival in the conviction that the *ontological capacity* for freedom, inseparable as it is from man's being, cannot really be extinguished, only temporarily banished from the public realm. [...] Given this faith, we have reason to hope that, as long as there are *human beings* who survive, the image of God will continue to exist along with them and will wait in concealment for its new hour. (*MM*: 112)

With this remark – bleak partly *because* of its collapse into eschatology – Jonas's final word on authoritarianism still fails to satisfy. Indeed, it very much conforms to Voegelin's observation that Gnostic political movements, in arguing for a truth revealed only to the few, are by necessity led to 'repress the truth of the soul' (1952: 165).

The following question now presents itself to us. If, as we have suggested, Jonas initially regarded authoritarianism as our best chance of survival, does it follow that his underlying theoretical framework – the ethic of responsibility, heuristic of fear and precautionary principle – is also objectionable, his thought having revealed its true face? Richard Wolin argues as much in his book *Heidegger's Children*, an interpretation we shall briefly consider as an alternative to Dinneen's.

Wolin provocatively argues that Jonas's ethical, political and metaphysical commitments place him in the tradition of *Lebensphilosophie*: the life-oriented school of German thought that flourished in the interwar years, and whose representatives include Oswald Spengler, Ludwig Klages, Ernst Jünger and – at least in some respects – Heidegger himself. One immediately notes a political commonality among these thinkers in their proximity to fascism, either through intellectual association, appropriation or, as in Heidegger's case, active participation. The explanation for this is a shared hostility to the modern epoch. According to the *Lebensphilosophen*, modernity's rejection of life, quality and soul in favour of materialism, equality and mind has led to the spiritual decay of the West. The charge levelled by Wolin is that Jonas's analysis of modern gnosticism amounts to more or less the same.

In support of this claim Wolin cites Jonas's ethical foundations, which, as we saw, involve a recognition of living beings as ends-in-themselves, in turn accounting for our responsibility towards them for their own sake. For Wolin, a 'risk entailed by Jonas's insistence on life as an absolute value is that our conception of the human good is devalued. Instead of setting our sights high and aiming at a notion of the good [...] Jonas's metaphysical vitalism tends to privilege "mere life" or survival' ((2001) 2015: 121). According to Wolin, this 'quasi-Darwinian' tendency explains Jonas's acquiescence to authoritarianism, preserving life even at the cost of the *good* life ((2001) 2015: 124). The underlying motivation is, he says, a 'resolutely antimodern epistemological orientation' and 'a disconsolate, Spenglerian sensibility' ((2001) 2015: 125, 129). As final proof Wolin points to an interview in which Jonas rhetorically asks: 'Was modernity perhaps a mistake that needs to be corrected? Are we on the right path with this combination of scientific/technological progress and increased individual freedom? Has the modern age put us in certain respects on the wrong track, which must not be pursued further?' (*CBE*: 26). Jonas offers no answer to these questions, but a critic might be able to divine one from his heuristic of fear. If we apply the latter to the trajectory of technological civilization as a whole, then presumably the entire modern epoch, given that it eventually gave rise to the ecological crisis, was indeed an error that ought to have been avoided.

However, Wolin's charge of anti-modernism only holds if one ignores key moments in Jonas's thinking. Yes, he bases his politics on his ethics, and his ethics emerge from his philosophy of life. But one can only arrive at Wolin's conclusion by overlooking the all-important idea of Man as a moral being, which tells us both *why* and *how* humanity must continue to be. The overriding importance of this duty is precisely why his ethics remains humanistic. Then there is Jonas's heuristic of fear, which does not, in fact, entail that modernity was an error. The heuristic requires that we extrapolate 'from *presently recognizable* trends in the technologic-industrial process'. This means that action to avert the ecological crisis should have been taken from roughly the 1960s, and our great sin is in failing to do so *from that point on*. It goes without saying that Jonas does not believe early modern Europeans should have envisaged, and acted to prevent, the ecological destruction which would eventually follow from the scientific revolution, since the former was not at that time conceivable as an empirically informed prediction. It is true that Jonas suggested we not 'be too modern' (*SE*: 20), but this was precisely a matter of rejecting 'certain developments which are ominous, which are dangerous, or which are undesirable', *not* that 'modernity as such was somehow a mistake' (*OR*: 3).

Finally there is the charge of a reactionary politics. It certainly cannot be denied that Jonas's theory of the state leads to an authoritarian conclusion, and to that extent I share Wolin's distaste. But it is not the case, as Wolin claims, that social democracy 'fails to make an impression' on Jonas – quite the opposite ((2001) 2015: 126). And we must also note that Jonas's political thought owes nothing to a social-Darwinian hierarchy of the fittest, a theory of 'natural characterology', or indeed any comparison with the natural world of the sort favoured by the *Lebensphilosophen*. On the contrary, he discusses at length the overwhelming *dissimilarities* between non-human life and the strictly symbolic and open-ended constitution of human society (*OF* II: 22–41). Therefore Jonas's theory of the state, however objectionable, does not resemble that of the *Lebensphilosophen* in the way that Wolin rather outrageously suggests.

V Freedom and the republic

One might ask: How is Dinneen right to say that Jonas's ethic of responsibility, heuristic of fear and precautionary principle are of genuine and ongoing value, at the same time as Wolin is right that Jonas's theory of the state is authoritarian, given that the latter is meant to follow from the former? The answer is that both are wrong to think that Jonas's theory of the state follows from his ethic of responsibility and philosophy of life. As we have seen, this is also an error Jonas himself makes, misapplying notions from the ethical and metaphysical spheres to that of politics.

Firstly there is the solely positive account of freedom that Jonas uses to characterize life. Already inaccurate as a description of the freedom of organismic being, when carried over to the domain of human political activity it also offers no basis to resist authoritarianism. Secondly, as discussed earlier, Jonas's comparison of the statesman's responsibility with that of a parent for their child all too easily collapses into paternalism. This is despite the fact that Jonas's argument for paternalism is actually at odds with his theory of parental responsibility, as the latter has 'to see that responsibility itself is not stifled, whether from its source within or from constraints without' (*IR*: 107). Of course, Jonas himself argues for the latter kind of restriction when suggesting that freedom might be banished from the public realm, as the ability to freely deliberate and shape our collective existence is the political counterpart to our being morally responsible beings.

The latter insight forms the core of the civic republican tradition, which has its origins in the Greek *polis* and Roman *res publica*. Recall that for the Greeks and Romans the common good was to be realized in this world, via active citizenship, rather than a utopia to come. Taking our cue from this original anti-gnostic form of politics, we shall now attempt to give like expression to Jonas's categorical imperative. Indeed, not only does the republican form of freedom follow from the responsibility engendered between equals discussed in Chapter 5, but a civic republicanism can even find textual support in Jonas's post-*Imperative* work. As he retreated from viewing authoritarianism as our best hope for survival, Jonas began to sketch out something like an ecological republicanism as the ideal political embodiment of responsibility.[13] When we synthesize his various suggestions and asides, we shall see that it offers a sadly incomplete alternative to his earlier theory of the state, and a way of redeeming Jonas as an environmental political theorist.

We mentioned at the start of this chapter that Jonas understood the political realm to be a manifestation of humanity's existential freedom: one that did not by itself recommend any politically substantive form of freedom. In search of such a theory, however, we might draw on his discussion of the Athenian city-state. In line with his philosophy of life, Jonas characterizes the freedom afforded by the *polis* as still, in one sense, *needful*, since 'to be a citizen of the Greek *polis* means to be a lawgiver and to be a lawgiver means to institute orders that bind others as well as oneself': therefore 'this is a freedom which acknowledges voluntary restrictions' (*PF*: 263). Of course, in ancient Athens the status of citizen was restricted to propertied, slave-owning men. The slave, by contrast, is unfree since 'deprived of making use of [his] will through the overpowering condition of a social order in which the sanctions imposed on his opposing the will of his master are overwhelming' (*PF*: 257). Both the full freedom of the Greek citizen and the captivity of the slave are therefore to be understood as relational, a 'power-condition' that is 'embodied in a legal order' devised by the former group (*PF*: 257).

This depiction of freedom and its opposite is neither purely positive nor strictly negative, but is instead closer to the republican variety: freedom understood as non-domination. According to Philip Pettit, republican freedom 'is negative to the extent that it requires the absence of domination by others' and 'positive to the extent that, at least in one respect, it needs something more than the absence of interference; it requires *security* against [...] interference on an arbitrary basis' (1997: 51; emphasis added). What provides this security? It is, Pettit says, the legal order that citizens form through deliberation. In both respects Jonas's presentation of the *polis* broadly aligns with this definition of freedom: the slave is unfree not because they are interfered with per se, but because they are dominated – interfered with totally and arbitrarily – whereas the citizen's freedom from domination is secured by the legal order they participate in constituting. The negative aspect is the absence of domination, and the positive the active constitution of the legal order that secures that non-domination. It goes without saying that the restriction of citizenship to propertied males is reprehensible, but this Classical feature is by no means inherent to the civic republican tradition. On the contrary, in seventeenth- and eighteenth-century Europe and North America the republican ideal was in the process of being extended to other social groups, before finding itself superseded by liberalism in the nineteenth century.

Jonas was initially highly sceptical of the possibility of recovering the Greek *polis* or Roman *res publica* for our times. Speaking of the nostalgia for antiquity characterizing the work of three friends and fellow émigrés – Arendt, Voegelin and Leo Strauss – he says: 'To be sure, the memory of those times [...] is [...] essential for our not getting lost in the necessities and compulsions and pushes of our modern age, which certainly has the danger of estranging us entirely from these eternal origins. But to hark back to them as a still available option is an anachronism, an escapism' (*OR*: 3).

Nevertheless, Jonas later drew on the republican ideal of citizenship as a way to cultivate new ecologically appropriate virtues. Speaking shortly before the end of his life, Jonas says:

> This is the one thing that keeps alive in me a modest hope [...]: that the sustained reflexion on the human good, on what is a worthy life for man, individually and collectively, and what we owe to it, may (with the help of some hard lessons) generate an internal tribunal of common conscience and good taste, to which even the noncaring must pay some obeisance, because

too blatant a transgression of its norms would incur the censure or revulsion of one's fellow citizens. (CR: 217)

We have here the familiar invocation of 'hard lessons', presumably of the ecological variety, generating knowledge in accordance with the heuristic of fear. But we also have a reference to citizens' *collective* cultivation of the good, which he elsewhere suggests might then be 'raised by the power of custom to a social norm' (*TME*: 75).[14] Drawing on other late remarks we shall attempt to develop this line of thought.

Responsibility can morally underpin both freedom from domination and the cultivation of ecological virtues that facilitate it. However, this would not be responsibility of the private kind – the parent–child paradigm – that proved so problematic when transposed to the public sphere. Instead we would point to a different sort of responsibility, discussed in Chapter 5: responsibility among equals. This form of responsibility, that of the community for each individual, follows from the limitations, vulnerabilities and dependencies constitutive of the human condition, interpreted through our moral tradition. If our social and symbolic existence is the ontological ground of the *polis*, it is the 'nonautarky of man', as Jonas puts it, which makes the political realm *necessary* (IR: 98). And our moral tradition, in recognition of this fact, instils in us a *public* responsibility to care for one another precisely because each of us belongs to the wider group that has greater power than any one member. This is, for example, the moral foundation of collective care for those living in hardship (economic or otherwise). The subjective aspect of this responsibility is known to us from the all-too-frequent encounter – at least on the streets of British cities – with homelessness and destitution. The call of responsibility is made of us less as an *individual* – since few of us can alone fully relieve others from poverty – but is rather made of us as a *member of the community*, because before us is a vulnerable good within our collective power to protect.

As stated, this alternative form of responsibility might underpin the republican kind of freedom. In order to be positively free citizenship involves participating in the establishment of the legal order, but to do so on an equal basis requires freedom from domination. As indicated, one example that ought to be guaranteed is support for vulnerable citizens: both freeing them *from* the domination of destitution and freeing them *to* participate in public life. More fundamental, however, are obligations to safeguard the existence and integrity of the *polis* itself, for without this, of course, it cannot afford freedom to each member. In antiquity such safeguards included personal and property rights, *provocatio* (in the Roman Republic, at least), military service and a range of other institutions to protect citizens and the *polis* from internal disorder and external threats. Whichever of these we deemed necessary to preserve, contemporary republicanism would now also have to consider the ecological *sustainability* of the *polis*, since its bare existence is imperilled through our own industrial activity. In other words, our responsibility to ensure the survival of the *polis* for the sake of each member means collectively imposing limits on economic development and industrial activity. We have here a public manifestation of responsibility for the idea of Man, but a call made to all of us as citizens, rather than just the paternalistic statesman.

Jonas makes three suggestions regarding institutions that might cultivate and formalize the norms guaranteeing such sustainability. The first is education. In *The Imperative* Jonas had characterized education as the space where private and public responsibility coincided, insofar as both parent and state are responsible for the child's future. The republican alternative would concur, but only on the basis of its different understanding of *public* responsibility. According to the paternalistic model of governance Jonas advocated, education, to the degree that it is a public responsibility, would presumably be an imposition of statesmen. According to the republican alternative, however, education would be guided by the responsibility of equals, thereby aiming to *develop equal citizens* who may partake in the collective life of the *polis*. As to the content of such an education, Jonas gives us a clue:

> Through education, through the way we bring up our youngsters and inculcate a style of life, we can have an influence on the forming of our consumption habits and make a certain frugality, a greater modesty, part of the social climate – or, [to] put it the other way, impose some penalty of shame, a social blemish[,] on excessive and vulgar hedonism [and] consumerism. (*CR*: 217)

This appears to refer to education's capacity to cultivate virtue and discourage vice, one that it already exercises, but now with the end of upholding our new ecological responsibilities. I will briefly attempt to justify this claim.

According to Aristotle, the virtues are character traits that, along with material and other circumstantial factors, contribute to flourishing (*eudaimonia*). An uncontroversial list of such virtues – to those belonging to the Western tradition – would include courage, honesty, benevolence and wisdom. The vices play an opposing role, diminishing *eudaimonia* through either an excess or deficiency of a particular trait: the virtue of courage, for instance, occupies the ideal middle ground between recklessness and cowardice. Now, the cultivation of virtue is not a merely private concern. According to Aristotle, politics aims at the common good: either that which is good for all citizens, or a good that can only be achieved by the collective. More specifically, public education – as well as private – in part concerns morality. In addition to teaching rules of conduct, moral education aims to cultivate virtue and discourage vice, drawing on a moral tradition for its content. Jonas's hope is that collective and sustained reflection on our new responsibilities might determine new ecological virtues and vices, which would then be incorporated into education. The core virtues he identifies are 'caution', 'frugality' and 'modesty' – all ancient, forgotten and now in need of a revival – to which we might add respect for life (*TME*: 67). The vices he points to are vulgar hedonism and consumerism, which, preventing us from carrying out our duties, would accordingly be a source of shame (*TME*: 69). Through this derivation of newly appropriate virtues and vices, citizens would better carry out their responsibilities: thus Jonas's ethics once again demonstrates its Kantianism.

The second institution we could point to is law, deriving moral justification from the norms just mentioned. Here Jonas explicitly takes inspiration from antiquity:

> [F]reedom can exist only if it limits itself. The unlimited freedom of the individual destroys itself because it is incompatible with the freedoms of the many [...]. In ancient Rome, for example, there were laws limiting private ostentation. Elected censors had the right to investigate whether displays of luxury were excessive. [...] This was a major infringement of personal freedom, but it was done specifically in the name of a self-governing citizenry. (*CBE*: 25–6)

Building on our discussion so far, we may say that although limitations on personal consumption clearly violate negative freedom from interference, they are nevertheless compatible with republican freedom from domination. If the citizenship endorsed legislation to limit consumption, for example, this would be justified by conforming to the aforementioned virtues that increase the ecological sustainability of the *polis*, which in turn helps safeguard the existence and essence of humanity.

One might object that there is a danger here in that *any* public intrusion in the sphere of the individual could then be justified in the name of the *polis* and the idea of Man. This is a real worry, to be sure, reminiscent of an objection often levelled at Rousseau's republicanism: if the general will is sovereign, then the individual may succumb to a tyranny of the majority. As Pettit argues, a republican government of the sort advocated by Rousseau would indeed be at risk of becoming 'a law unto itself' and making the individual 'vulnerable both in relation to the state and in relation to our fellow citizens' (2012: 24–5). This problem is not inherent to republicanism in general, however. The Anglo-American strand – Locke, Harrington, Madison and Jefferson – stressed the kinds of institutional safeguards that protect core individual freedoms and ensure contestability of legislation, thereby mitigating an otherwise legitimate concern.

The final suggestion that may be extracted from Jonas's later work pertains to international law and governance. Tackling the ecological crisis is predominantly the responsibility of globalized industrial nations, and so coordination of this effort will have to be – at least to some extent – an international effort. In *The Imperative* Jonas had envisioned a global government as best able to rise to this challenge, but given that the alternative we are here developing takes inspiration from the *polis* or *res publica* – that is, the city state, or at most the nation state – it appears to be incompatible with operating on such a scale. It is true that the Roman Republic, even before its transition into the empire, stretched across the Mediterranean, and was therefore not geographically confined in the same way as the Athenian city state. But this expansion was, of course, achieved through military subjugation, and is therefore at odds with the modern and inclusive model of republicanism under discussion.

Jonas suggests, therefore, that the solution must be the political creation of a 'peaceably united humanity' (*TME*: 71). This would have to be achieved through international law and treaties, as 'bodies must be established that address these [global] issues and enjoy a sort of international authority that governments and corporations cannot easily escape. [...] This would be a step on the road to a real *cosmopolis*' (*FWT*: 118). Through such a system – a United Nations with real authority, perhaps – the *polis* would be situated within a *cosmopolis*, and politics could thereby legislate on the global workings of technological civilization without recourse to authoritarianism. To be sure, this Kantian goal may well be utopian: Jonas does not offer much detail as to how it might be achieved, and indeed, 'a mysterious evolution of mankind toward peace and world order' is cited by Voegelin in his list of gnostic utopias (1952: 172). In a public address given a year before his death Jonas concedes as much, but with the mere possibility of an internationalist solution to the threat of ecological ruin says that he is 'more hopeful than at the point when, fifteen years ago, I published my book *Das Prinzip Verantwortung*' (*FWT*: 119). On an uncharacteristically optimistic note, then, we conclude our attempt at devising an alternative political theory from Jonas's late work.

Jonas's political philosophy is quite clearly the most objectionable component of his philosophical system. This is demonstrated not only by the criticisms levelled at his theory in *The Imperative* but also by the fact that an attempt to find an alternative, in civic republicanism, could only draw on suggestive remarks and asides. And yet Jonas's contributions as a political theorist are not without merit. His rejection of the liberal assumption of neutrality between competing conceptions of the good, though controversial, is defensible given that the moral content he suggests is responsibility for the idea of Man and life on earth: duties to which no one could reasonably object. Then there was his heuristic of fear and associated precautionary principle, the latter of which has even been taken up, albeit in weaker forms, by the United Nations and European Union.

Finally, we sought to develop Jonas's sketch for a republican alternative to authoritarianism that could justify productive and consumptive constraints. It might be argued that this attempt recasts Jonas as akin to those thinkers he mocked for nostalgically harking back to antiquity: Arendt, Voegelin and Strauss. But in a sense Jonas's political philosophy as presented in *The Imperative* already conforms to this type. The only difference is that rather than reviving the highest aspects of Classical political theory, Jonas there draws on the most dangerous. Intentionally or not, he recalls in particular the dictatorial powers that the Roman Republic would temporarily grant to a chosen magistrate. In times of crisis the Senate would suspend democracy for six months or until the danger to the republic, either internal or external, had been dealt with. Jonas's call for a suspension of democracy to confront the ecological crisis – a suspension that he later specified should last only as long as is necessary – is curiously reminiscent of this constitutional practice. By contrast, the attempt to develop an ecological republicanism from other aspects of his thought neatly aligns with the normal conditions of the Roman Republic: a clearly preferable source of anti-gnostic inspiration. Most importantly, it conforms to a principle that Jonas himself advocated: '[T]o keep watch over the humaneness of the measures by means of which we are trying to avert catastrophe. For these measures could be such that the whole thing we are trying to save goes to the devil' (*CBE*: 29).

7

Towards a richer bioethics

I The dignity of the person

Of the novel ethical problems posed by technological civilization we have so far focused on those relating to the environment, and analysed Jonas's theory of responsibility accordingly. This is true to his thought insofar as establishing the norms and institutions that might prevent ecological catastrophe was the central task that he ascribed to moral and political philosophy in the present century (*MM*: 51). But Jonas's ethics has a second practical dimension, pertaining to biotechnology and the life sciences, which is our topic in this final chapter.

The importance of bioethics to Jonas's thought is clear from the reference in *The Imperative* to an envisioned 'applied' counterpart dealing with such issues (*IR*: 21). Jonas's subsequent volume, *Technik, Medizin und Ethik*, explicitly fulfils that promise (*TME*: 9), collecting the majority of his essays on bioethical issues. In this chapter we shall see how these writings allow us to think through novel problems posed by developments in biotechnology and the life sciences, and how this effort relates to his critique of gnosticism and theory of responsibility for the idea of Man. In particular, we shall see that Jonas was an early and perceptive critic of what has since become known as transhumanism: the drive to alter the human condition itself through biotechnology.

Firstly, however, we shall introduce Jonas's somewhat idiosyncratic approach to these issues. According to Albert Jonsen, 'Jonas was the first philosopher of eminence to arrive on the medical ethics scene,' but his influence was limited thanks to a style which 'was, perhaps, too ontological and conservative for the typical American ethicist' (1998: 77).[1] If accurate, this is less a reflection of Jonas's significance than of the lamentable reductionism that prevails in Anglophone bioethics: Jonas's arguments do indeed draw on his wider ethical and metaphysical thought, and this undoubtedly limits their appeal in a discipline oriented towards instrumental solutions. Less defensibly, it must be admitted that on occasion the force of Jonas's rhetoric – as distinct from the substance of his arguments – can be dogmatic. In general, however, his thought embodies the nuanced approach that is demanded by the central context in which bioethical issues arise: healthcare. Healthcare has an inherent normative thrust – curing disease – that marks it out from *technē* as such. Although medical developments cannot be simply affirmed without the risk of committing wrongs both great and small, neither are we usually able to issue simple prohibitions. Instead 'we find an area of fluid boundaries, subtle value judgments, and controversial decisions' (*MM*: 50).

At this point one might ask exactly *why* philosophers such as Jonas may make prescriptions for scientific and technological research, when the arts and humanities themselves are subject to no such restrictions. The difference, of course, is that the latter belong exclusively to the domain of speech and ideas, whereas the former also belong to the realm of action. As we saw in Chapter 1, modern natural science is distinguished from its pre-modern forebear by emphasizing experimentation. For this reason it is inseparably tied to practice, a connection that is only compounded by its natural application to technology. While freedom of *expression* is conceivably an absolute right, there is no equivalent right to freedom of *action*, and so we must regulate the latter in line with moral norms (*SB*: 255–8). The ethical issues underpinning such regulation include, least controversially, material and personal harms captured by a consequentialist framework. Jonas's central concern, however, is that even with a utilitarian sanction the will to medically assist may come into 'conflict with human dignity' – that is, the personal dignity that humans alone possess courtesy of our moral being (*MM*: 50). Even if moral philosophy, in response to such cases, 'has nothing to offer except compromises between conflicting principles', it remains a responsibility to think through such developments and ensure that the highest good is not sacrificed in the name of amelioration (*MM*: 50).

How, then, are such issues to be adjudicated? From where do we derive our notion of human dignity, and how is it meaningfully substantiated? Jonas's answer to the latter question is the heuristic of fear and ethic of responsibility. We recall from Chapter 5 that the ultimate object of our responsibility is the idea of Man as a moral being, requiring, by necessity, the presence of human life as its bearer. This alone is enough to account for a responsibility to ensure the existence of future generations, but in the realm of the life sciences there are few situations in which this is a plausible concern.[2] The threats to the idea of Man that typically follow from developments in biotechnology and the life sciences are more ambivalent, requiring the kind of deliberation mentioned. It is in this connection that Jonas's heuristic of fear has another crucial role to play. We recall from Chapter 6 that the heuristic's function was to help us avert courses of action in which the existence or essence of human life were threatened. As stated, in the cases of humanity's bare existence or its capacity for morality the nature of the threat is clear. But in those instances where the human essence might be compromised or violated, the fear *itself* helps us to identify and better understand precisely what is at stake. As Jonas puts it: '[W]e need the threat to the image of man – and rather specific kinds of threat – to assure ourselves of his true image by the very recoil from these threats' (*IR*: 26–7; emphasis removed).

Jonas's logic here might sound circular, presupposing that which is subsequently discovered: as though we imagine a threat to the idea of Man, but only through our fearful response to that threat do we discover the idea of Man. However, while Jonas admits that his argument has an air of paradox about it, it is not, in fact, circular in the sense just given. What Jonas means is the following: the *demands* made of us by human dignity – the dignity that follows, we recall, from our moral being – only reveal themselves when we perceive or imagine a violation of that dignity. With the knowledge of these new demands, the *eidos* of humanity is further expanded and refined, and so we come to '*know the thing at stake only when we know that it is at stake*' (*IR*: 27).

A little more has to be said in justification of this claim. Human dignity, the ethical expression of the idea of Man, functions in this oblique way because dignity in general is not the kind of thing that can be comprehensively described independently of circumstances. It is not an object, but a status, denoting both the moral significance of a being *and* the particular ways in which it can and cannot be treated. While the former is rationally demonstrable – as we saw in Chapter 5 – the latter can only be discovered contextually. And the question of context is here paramount: historically, the stable temporal and spatial reach of our actions broadly aligned with time-honoured norms governing interpersonal relations. But since emerging technologies and novel scientific practices provide us with unprecedented ways of relating to fellow persons, we have to draw on observation and imagination to discover which of these, if any, violate our dignity.

This brings us to the procedural question of how bioethics should be conducted and fed into legislation. Here Jonas's suggestions notably contrast with his earlier application of the heuristic of fear. Unlike the ecological crisis, which is a matter of quantifiable threats to the existence of humanity, threats to human dignity are qualitative, and less permitting of expertise. While one can be an expert in climate science, it is not clear how one could be an expert in perceiving violations of human dignity – anyone with a normal sense of morality would count as such, entailing that the claim to 'expertise' is meaningless. Something similar is true of bioethical cases where moral absolutes are not in play, and one must weigh concerns such as integrity, non-personal dignity and utility against one another. While the bioethicist's training might assist them in avoiding errors of reasoning, there is no objectively correct balance to be struck between these values: we must instead decide which matters most to us, with a view to the kind of society we wish to be.

Since there are no moral experts, and bioethical issues themselves concern the 'extra-scientific sphere and wider society', Jonas suggests that such decisions be made by bodies 'constituted by laymen from all walks of life' (*TME*: 79). The advantage of such an approach is that it may better avoid 'the danger of subjective arbitrariness' present whenever we deal with qualitative properties that 'only become apparent in personal perception' (*TME*: 86). The following reflections, then, should be understood not as a claim to moral expertise but rather as an attempt to articulate concerns underpinned by fundamental elements of our moral tradition. We shall start with one of Jonas's most influential contributions to bioethics: analysing the ways in which medical research on human subjects can violate personal dignity.

II Human beings as means

One of the main reasons for the development of bioethics as a discipline was an increased awareness of ethical violations taking place in the field of medical research. Although the most heinous crimes occurred in the concentration camps of Nazi Germany and the POW camps of Imperial Japan – with similarly horrifying reports emerging from North Korea today – ethical scandals were by no means restricted to totalitarian states. The Tuskegee Syphilis Experiment run by the United States Public Health Service from 1932 to 1972 is a notorious case in point. Six hundred impoverished African American participants were enrolled in the study to observe the effects of untreated syphilis: 399 of the participants already had the disease, while the remainder acted as healthy controls. However, none of the infected actually knew they had syphilis: they were told only that they were under observation for 'bad blood', and even after penicillin became an established treatment for the disease in the mid-1940s none were cured. As a result, dozens of participants died, forty sexual partners contracted syphilis and nineteen children were born with its congenital form.

The Tuskegee Experiment is a stark example of the kinds of wrongdoing that concern Jonas: wrongs that a purely consequentialist approach fails to adequately capture. Clearly, one can make a strong utilitarian case against the great physical harm done to the participants and their dependents, and the psychological distress of subsequently discovering that from a certain point in time the former was entirely preventable. But a strict utilitarian could also argue that these harms, since pertaining only to a few hundred individuals, were outweighed by society's interest in developing a cure for the disease. As the benefits of the experiment extend from the contemporaries of those who suffered to the generations following who enjoy freedom from syphilis as a result, on a simple balance of benefits and harms the result is surely a net positive.

Few, if any, bioethicists would actually make such a case – although in the discipline's infancy questionable appeals to the 'greater good' were not unheard of.[3] And, clearly, any such defence of the Tuskegee Experiments, or experimentation broadly like it, would rely on a number of empirical assumptions. The two most important are, firstly, that there is a direct causal connection between the experiment and the discovery of a cure, and secondly, that other citizens, both present and future, do not find the experiment so appalling that any gain in utility derived from it is thereby outweighed. Were neither of these conditions to hold then the utilitarian could surely not support such experiments. But the mere fact that their moral assessment would turn on outcomes, rather than the violation of basic rules of conduct, is precisely what is so concerning. The appalling treatment of participants in the Tuskegee Experiment can only be truly captured by a different ethical vocabulary: that of human dignity. Jonas's seminal essay 'Philosophical Reflections on Experimenting with Human Subjects' uses the concept to identify the limits of what can rightly be done to individuals in experimental research, and in so doing further illuminates the idea of Man.

To begin with Jonas posits that a sacrifice made for medical research be exactly that: a *sacrifice*, which 'must be absolutely free' (*PE*: 111). In the bioethical literature this is typically characterized as giving one's informed consent to be used in research. Clearly, in the Tuskegee Experiments the lack of informed consent – leading to the unknowing infection of sexual partners and the resulting children – is one of the main reasons why it was so immoral. But even informed consent, if left unsubstantiated, is insufficient (*SB*: 260–1). For if the ill and impoverished participants of the Tuskegee Experiment had given their informed consent in exchange for some sort of financial compensation, say, then we might justifiably think that they had been taken advantage of. In order to avoid such exploitation in the recruitment of research participants, thereby ensuring that informed consent is truly free, Jonas offers some robust criteria. He suggests that we prioritize, and progressively work down from, those volunteers who simultaneously (a) least need remuneration, (b) have the best understanding of the experiment and the risks involved and (c) most believe in, or identify with, the purpose of the research. Not only does this totally reverse the 'availability and expendability' logic of the Tuskegee researchers, it also makes the researchers *themselves* the most suitable candidates for participation (*PE*: 124). Impractical though this may be, 'with all its counter-utility and seeming "wastefulness", we feel a rightness about it' (*PE*: 125).

Jonas insists that such free and informed consent is a 'minimum requirement' of medical research, and the reason why, as indicated, is that it preserves the human dignity of those participating in the experiment (*PE*: 121). In making the free decision to partake for the sake of the cause itself one acts as a full moral agent. In a Kantian vernacular, we might say that by consenting in this way the participant remains an end while also being treated as a means, without doing so they are instrumentalized as a means only. Making this very point, Jonas claims that '[o]nly genuine authenticity of volunteering can possibly redeem the condition of "thinghood" to which the subject submits' in an experimental situation (*PE*: 109).

There is an ambivalence in Jonas's argument, however, which we must address. On the one hand, he claims that a robust definition of informed consent is necessary to avoid a violation of human dignity in the experimental situation. On the other hand, when drawing up his criteria for how to ensure free and informed consent, Jonas suggests that they are ideals, not absolutes: '[A] descending order of permissibility leads to greater abundance and ease of supply, whose use should become proportionately more hesitant as the exculpating criteria are relaxed' (*PE*: 123). This represents a concession to pragmatism on Jonas's part, since he acknowledges that strictly observing his criteria would suffice 'neither in numbers nor in variety of material' for statistically meaningful experiments (*PE*: 122).

Could Jonas's criteria for free and informed consent be relaxed without treating participants merely as means, and if so, how? We can imagine a degree to which two of the three could be loosened: we might allow for someone who did not grasp the finer points of the research project in which they participated, as long as they fully understood its potential risks, and someone who identified with the purpose of the research not to the extent of the researchers themselves, perhaps, but nevertheless such that they felt it to be a worthy cause. Much more troubling would be relaxing the criterion of recruiting those who least need material remuneration, which was included to prevent exploitation of the socio-economically vulnerable. So even if we do not accept a wholesale weakening of Jonas's criteria, we can, then, allow that free and informed consent entails a belief in the purpose of the research, comprehensive understanding of its risks and a lack of socio-economic vulnerability. Needless to say, testing for these would be difficult in practice – but this is no fundamental objection to their rightness.

We have here, then, an example of how the demands of human dignity emerge from a concrete situation: in an experiment our dignity is violated if we are treated as a thing rather than a person, revealing the ethical necessity of free and informed consent, which in turn renders Tuskegee-style experiments impermissible in principle. However, proceeding in this way implies that the burden of proof lies on the side of the individual, and that the societal demands made of us in this domain are legitimate unless shown to be otherwise. But why, Jonas asks, should we assume this to be so? For it would only hold if the medical-scientific enterprise bestowed researchers with overwhelmingly strong grounds to make demands of individuals, and it is not self-evident that it does. Why should we not think the opposite: that science shoulders the burden of proof in terms of what it may ask of the individual?

To be sure, Jonas accepts that healthcare is a generic good, but it does not follow from this that we have obligations to contribute to its progress. Should we refuse to volunteer for experiments, such that no novel research took place, society would not collapse. On the contrary, '[i]f cancer, heart disease, and other organic noncontagious ills, especially those tending to strike the old more than the young, continue to exact their toll at the normal rate of incidence (including the toll of private anguish and misery), society can go on flourishing in every way' (*PE*: 117). There is, to reiterate, no obligation to contribute to medical research for *society*'s sake: it will be just fine. What may rightly provoke a self-sacrifice – either of possessions, such as time and money, or perhaps even a part of one's bodily self – is the vulnerability of *individuals* suffering from the diseases. Even this must not be understood as an obligation, however. There may exist an absolute duty to help others if it is within one's power to do so and the risks are not prohibitively great. But the link between participation in medical research and saving lives is not straightforward: the former only potentially increases the chances of the latter, and to a highly variable degree. So although participation might be commendable – an imperfect duty, as Kant would say – it would not seem to amount to an obligation.

Jonas acknowledges that there are limits to this argument, however. His rejection of duties to society rests on the observation that it will continue to exist without any contribution to the progress of medical science. But there are, at least hypothetically, circumstances where this will not be the case, such as the need for a cure to combat a deadly pandemic – one closer to the Black Death of the fourteenth century than the coronavirus outbreak of 2019–20 in terms of severity. Can we then be said to have duties to society? And what role does human dignity then play?

Himself a veteran of two conflicts, Jonas suggests that there may be a parallel between such a situation and a military emergency. He suggests that in wartime 'society itself supersedes the nice balance of the social contract with an almost absolute precedence of public necessities over individual rights [...] a near-totalitarian, quasi-communist state of affairs is *temporarily* permitted to prevail' (*PE*: 115). This claim is too broad if meant for just any conflict, since not all wars imperil society, but we would be willing to concede that intrusions into the private realm, up to and including military conscription, are permissible if necessary for the very survival of a just polity.[4] In this case there are indeed similarities. Prevention of both the military and the medical emergency are good ends. Both might ultimately be to no avail: the war might not be won, the pandemic not averted. And finally, in both cases human beings are used as means: the soldier is trained and ordered to kill, the research participant submitted to experimentation. Are they then alike, the one as permissible as the other?

There might yet remain a subtle but significant difference that prevents us from concluding in the affirmative. Although the soldier must carry out their orders under pain of being court-martialled, refusal is always a possibility. And when the soldier does carry out their orders, they can remain an end while being treated as a means: the soldier can be brave or fearful, cunning or dim-witted, they can excel or be merely adequate. The conscripted experimental subject, by contrast, can be none of these things – or rather, while they might be otherwise virtuous people, they cannot be virtuous *qua* experimental subject. The research participant has no scope for agency and excellence because they are reduced to a mere *thing*, a source of data, while even in conscription the soldier retains a modicum of personhood and with it, human dignity, in their having some choice between actions and conduct. Although this distinction is very slight, it may allow us to differentiate between forms of conscription.

Finally Jonas considers the principle of reciprocity. Might we be indebted to those who have already participated in medical research, from which all of us benefit? Evidently those to whom we are supposedly indebted would have to still be alive: if they were not, then we cannot benefit them in return. But Jonas is sceptical that we have a duty of reciprocity even to those from whom we have benefited and who, being alive, we may still benefit in return. The reason why is that, as discussed, a true sacrifice must be voluntary: it is from this that it derives its goodness, above and beyond the utilitarian aspect. And of course, this very fact entails that one cannot be *expected* to make such a sacrifice in return, as it would then no longer truly be a sacrifice. Hence, as Jonas says, 'precedence must not be used as a social pressure' to extract consent where it is not genuinely forthcoming (*PE*: 120).

III The threshold of life and death

Jonas's analysis of free and informed consent contributes to an understanding of how we may and may not treat human beings in experimental situations. Of course, that consent is even a possibility presupposes that the human beings from whom it must issue are living and conscious persons. Jonas was also concerned, however, with new techniques for both the termination and artificial prolongation of life. These invite us to question what we owe to human beings who are not – or are at least no longer – conscious, and thus incapable of consenting to treatment. Of these bioethical issues abortion was arguably the most hotly contested, yet Jonas wrote surprisingly little about it.[5] He did, however, write insightfully on the ethics of euthanasia and the moral significance of demarcating death for the purposes of acquiring organs. Beginning with the latter, we shall see what else Jonas's theory of the idea of Man can tell us about the boundaries of appropriate treatment.

In 1968 the Harvard Medical School published a report arguing that irreversible coma should become the medical definition of death. Jonas wrote an ethical critique in response – 'Against the Stream' – which remains of philosophical significance for what it tells us about the conceptions of humanity presupposed in bioethical debates. This broader point emerges from Jonas's more immediate argument, which is against the motivation of the report's authors. For the purpose of the proposed redefinition of death as irreversible coma was not strictly scientific – that is, the best-informed theory available – but rather *instrumental*: to maximize the number of organs available for transplantation.

To be sure, there is a clear utilitarian rationale for such a redefinition, which ought to be given its due: people regularly die for want of available organs, organs that will never again be appreciated by their original possessor if he or she is in an irreversible coma. Both of these statements have persuasive force, and yet we feel uneasy when presented with them as a case for redefining death. The reason why is the aforementioned instrumentalism: the report is not concerned with whether the irreversibly comatose patient really *is* dead, but whether we should declare them dead so as to allow us to freely make use of them. It is the former issue alone which concerns medicine; the latter, in Jonas's words, instead conforms to 'the ruling pragmatism of our time which will let no ancient fear and trembling interfere with the relentless expanding of the realm of sheer thinghood and unrestricted utility' (*PE*: 142).

In response one might argue that this says nothing about the medical question of whether an irreversible coma is actually an inappropriate definition of death. For one could say that, yes, the motivation behind the report is deplorable, but this does not prove that the proposed definition of death is itself wrong. If it turned out to be correct, then the report would in fact be advocating not for the reduction of human beings, in the fullest sense, to repositories of useful tissues, but rather for *dead* human beings to become such things. And although this might still trouble us, it would not do so to anything like a comparable degree.

In formulating a reply to this line of argument, Jonas's philosophical biology and anthropology become relevant once again. The idea that a human being is dead once the higher functions of the brain are irreversibly lost betrays, he says, 'a curious remnant of the old soul–body dualism. Its new apparition is the dualism of brain and body' (*PE*: 140–1). For the notion implies that once certain faculties – peculiarly human ones, to be sure – are irretrievably lost, the *human being* is dead, or as good as. But why draw this very gnostic equivalence, when, as we saw in the Chapter 2, human beings are both mind *and* body, the latter referring to not only the brain but rather the organism as a whole? As Jonas says, '[I]dentity is the identity of the whole organism [...]. How else could a man love a woman and not merely her brains? How else could we lose ourselves in the aspect of a face? Be touched by the delicacy of a frame? It is this person's, and no one else's' (*PE*: 141).

Now, this argument is beset by a problem of its own, namely the ambiguity concerning when the organism as a whole is dead. Part of the discomfort with declaring an individual to be dead on the basis of irreversible coma – or even brain stem death, the current medical definition in the UK – is that they might still be circulating blood and breathing, even spontaneously so (i.e. without medical assistance). A breathing, moving, *metabolizing* being is by definition alive, even if permanently unconscious. However, if we follow the claim that an individual can only be dead once they have ceased to breathe and circulate blood, we are faced with the problem that hair and nails will continue to grow for a period thereafter. Here is evidence, however small, of the very metabolic activity by which Jonas characterizes life. Ought we then say that an individual is only dead once even these processes have ceased? It seems that consistency would demand it – yet Jonas argues that this is not, in fact, required by his philosophy of life. Rather, even if some *localized* metabolic activity continues after respiration, circulation and sensation have irrevocably ceased, with these latter cessations the organism understood *as an integrated whole* has died: '[T]he effect of their functioning, though performed by subsystems, extends through the total system and insures the functional preservation of its other parts' (*PE*: 137). His definition, then, ultimately relies on the organism as a teleologically organized being. We could note that, after all, the body as a whole will decay – surely evidence of death – while hair and nails continue to grow, but as long as the respiratory and circulatory systems are intact it will not.

Though this definition is imprecise, it is more satisfactory than those that rely on brain stem death or irreversible coma as their criterion. If the former lacks the clarity of the latter, then this may simply be truer to the phenomenon. As Jonas says, referencing Aristotle: 'Giving intrinsic vagueness its due is not being vague. [...] Reality of certain kinds – of which the life–death spectrum is perhaps one – may be imprecise in itself, or the knowledge obtainable of it may be. To acknowledge such a state of affairs is more adequate to it than a precise definition, which does violence to it' (*PE*: 136).

To return, then, to the issue with which we began the present section: What does this tell us about the permissibility of acquiring organs and other tissues from human beings? Certainly, it does not provide us with a *categorical* prohibition, for the following reason. If only human dignity represents an inviolable principle, but an organism permanently lacks the capacity for morality that denotes such dignity, then it would not enjoy its protection. And if organ harvesting from the conscious is an action prohibited by human dignity, then a human being either in an irreversible coma or whose brain stem has died would not possess the specific dignity violated by such an act. Even if not categorically prohibited, we may still account for the intuited wrongness of the act, however. Firstly, we could point to the utilitarian consideration of any distress and horror which might be caused by seeing loved ones put to death by the state, even if for organs which might save the lives of other citizens. Whether this concern would hedonically outweigh the gains is not obvious, but it is only right that we consider the emotional harms that such a policy could entail. A second objection follows from the fact that the being in question *is alive*, which is sufficient, even in the condition discussed, to be wronged. They may no longer be a person in a Kantian sense, but the organismic person is nonetheless still there. On this basis they possess integrity and non-personal dignity – the dignity of ends – both of which would be violated by killing for the harvesting of organs.

The ultimate reason for our objection to live organ harvesting, however – though admittedly flawed – is the following. It is the knowledge that *this* living being *was* a person in the Kantian sense, even if they are no more. There remains in the living non-conscious body an echo or trace of the full person that was there, and although they do not possess human dignity now, the mere fact that they *did* (or are presumed to have done) colours our perception of said body. However 'irrational' it may be, the brain-dead individual is still a 'he' or a 'she', not yet an 'it'. To put them to death solely to make use of their tissues is to instrumentalize a living being historically connected to personhood, and although this might not provide us with a rational categorical objection, it provides an emotionally compelling one. A critic might ask whether one could extend this sentiment to anything that bears the trace of the human being that was: a dead body, a skeleton, ashes, even medical waste. In fact, the suggestion is not as absurd as it perhaps first seems. We do treat the deceased with a degree of respect, after all: closing their eyes to appear at peace, cleaning the body and of course holding a funeral to pay our respects. This is not meant to imply that a dead body makes any strong ethical demands of us, only that it makes sense to say that it presents *some*, which weaken the further from the living being we eidetically move.

The foregoing discussion began with the question of when a person can be medically declared dead, with a view to establishing the permissibility of harvesting their organs. The proposal of the Harvard Medical School essentially amounted to non-consensual active euthanasia – killing patients who cannot assent to it – and this for an instrumental purpose.[6] Such a prospect amounted, we argued, to a violation of integrity, non-personal dignity and the lingering echo of their human dignity. Curiously, however, human dignity is later cited by Jonas as a justification for non-consensual *passive* euthanasia: that is, letting patients who cannot assent to it die.

Such a patient would have to be alive solely by virtue of medical intervention, and permanently incapable of regaining consciousness. The consent criterion might yet be fulfilled if the patient had previously expressed, in an advance notice, that under such circumstances they would want life support to be removed. But let us assume the patient has not done so. In that case, Jonas says that '[r]eason, sanctity, and humanity' suggest that 'the patient *ought* to be allowed to die; stoppage of the sustaining treatment should be mandatory, not just permitted' (*RD*: 35). The reason for this very strong claim is that 'something like a "right to die" can […] be construed on behalf and in defense of the past dignity of the person that the patient once was, and the memory of which is tainted by the degradation of such a "survival"' (*RD*: 35). This appears to be the very same appeal to the 'echo' of human dignity invoked above as a reason for not harvesting the organs of the permanently non-conscious. And yet Jonas seemingly invokes the principle to argue that the permanently non-conscious should be allowed to die as a matter of course. Is there then an inconsistency here?

Perhaps something subtler is in fact at play, involving the alignment or misalignment of the echo of human dignity with non-personal dignity and integrity. The issue is whether the entire organism – mind *and* body – has ceased to live spontaneously, or whether only one side has done so. To allow the permanently non-conscious patient to die through withdrawal of life support is to 'let the poor shadow of what was once a person die, as the body is ready to do, and end the degradation of its forced lingering' (*RD*: 35). The intuition relied upon here is that a life artificially perpetuated at only the most basic level is not one that befits their trace of human dignity: the entire person has ceased to live of its own accord, and thus permanently lacks non-personal dignity *and* human dignity. In such a situation, their echo of human dignity might well be better respected by withdrawing treatment and allowing death to take its course (we stop short, however, of concurring with Jonas's claim that it is obligatory). By contrast, when the body is alive of its own accord, still possessing non-personal dignity and integrity, then the trace of human dignity demands that we not actively end their life: one half of the psycho-physical personhood lives of its own accord. And of course, if the person remains conscious but survives only through life support, their full human dignity demands the artificial perpetuation of their organismic ends. Here we have an example of the subtlety of the demands of dignity, as befits the complexity of the issues at hand.

IV The future of the human condition

The final topic to be addressed in the present chapter is transhumanism: the drive to 'enhance' human beings through biotechnology. Transhumanism, or human enhancement, is advocated in various forms, depending on the faculty or capacity in question. Future humans might be physically enhanced, becoming fitter and stronger, or cognitively enhanced, with greater creative abilities and powers of recall; they might be morally enhanced, becoming more empathetic and just, or perhaps enhanced in lifespan – immortality being, according to the transhumanist John Harris, the 'Holy Grail of enhancement' (2007: 59). Harris's religious choice of metaphor is apt, as enhancement in general and immortality in particular have long featured as *desiderata* in religious and mythological texts. Indeed, contemporary transhumanism is, as we saw in Chapter 1, a kind of religion. As Bacon himself said, the eschaton of the scientific method is the mastery of nature, both human and non-human, with nothing to prevent this drive leading up to and including biological immortality.

What is decidedly new, however, is the rapid development of biotechnology in the last half-century, which has the potential to finally realize Bacon's dream of mastery over our biological constitution – or at least make aspects of it a plausible scientific prospect. Jonas was quick to recognize this, and his writings on genetic engineering as a means and transhumanism as an end remain vital guides to the ethics of both. We shall reconstruct his arguments, and in doing so resume our refinement of the idea of Man. But another concern is relevant here also: the future of the *human condition*, which is here understood in the sense outlined by Arendt (1958: 1–11). Human enhancement promises or threatens, depending on one's perspective, to transform both our freedom, as it manifests in our unique degree of existential openness, and our finitude. Both belong to the human condition, the value of which is revealed to us anew when contemplating its modification. The human condition does not carry the moral weight of personal, human dignity that belongs to us as moral beings, but we may instead appeal to our organismic non-personal dignity to explain its normative significance. As stated, the latter does not represent an inviolable boundary, and so wherever transhumanism threatens *only* this, and not also human dignity, we cannot categorically object to it. We may, however, follow Jonas in doubting its wisdom.

We shall look firstly at the key method discussed, namely, genetic engineering. In line with his heuristic of fear Jonas offers an 'existential critique' of genetic engineering, imagining what it would be like to be an engineered human being in order to reveal the problems that would be raised by the practice (*PE*: 165; emphasis removed). Although transhumanists frequently couch their arguments in terms of greater freedom – which means, for them, freedom from biological limits – Jonas's critique leads him to the conclusion that genetic engineering could in fact *compromise* our freedom. He makes two observations: firstly, that genetic engineering would undermine our existential freedom to self-develop, and secondly, that it would corrupt the relation between those who performed it and those who underwent it. These concerns pertain not to the physical consequences of manipulating the genome, which are not deterministic, but instead to the personal and social significance of such an intervention.

The first change Jonas points to is that the process of self-becoming is disrupted by the knowledge of having been manipulated in order to be a particular way. Regardless of whether that manipulation works as intended, the intention itself could change the person's self-understanding as they second-guess the authenticity of their physical or psychological being. As Jonas says, '[I]t does not matter one jot whether the [engineered] genotype is really, by its own force, a person's fate: it is *made* his fate' (*PE*: 163). What is lost is the spontaneity of becoming by instead measuring oneself against a pre-established design:

> [T]he sexually produced genotype is a novum in itself, unknown to all to begin with and still to reveal itself to owners and fellow men alike. Ignorance is here the precondition of freedom: the new throw of the dice has to discover itself in the guideless efforts of living its life for the first and only time, i.e., to *become* itself in meeting a world as unprepared for the newcomer as [the newcomer] is for himself. (*PE*: 161)

Jonas's argument makes, to reiterate, no claim as to the metaphysical possibility of freedom; at stake is a relational, *inter-subjective* freedom.

His existential objection arguably has three limitations, however. Firstly, it would appear to apply only to people who do not believe in a creator deity: for if one already holds that life is created in a particular way, one would presumably not then experience any additional loss of existential freedom in having been genetically engineered. Secondly, many people who *do* believe they were created in a particular way by God seem not to find this belief alienating, but rather a source of comfort. It might be the case, then, that knowledge of being genetically engineered would come to be perceived in the same welcome light.

To take the latter point first, one suspects that the religious believer's comfort is more to do with the specific author of creation than being created per se. It is, in other words, due to the fact that they think of themselves as created by a supernatural being, which provides a 'higher' reason for everything being as it is. Knowing that one is engineered by one's predecessors would be categorically different, since lacking this divine status. We can also reject the first possible objection. Being genetically engineered would still, we suggest, be experienced as a loss of existential freedom by those who already believe in a creator deity. Precisely *because* the religious assume a supernatural creator the *intent* behind creation is unknown: one discovers God's (supposed) plan for us in its actual unfolding. As such, the religious do not experience one's self-becoming against a known prior intention, as would surely be the case in genetic engineering.

The third problem with Jonas's existential objection, following from the response just given, is that it would only hold in the case of persons who actually knew they were engineered. This much is true. The transhumanist might then suggest that, if possible, we make enhancements either to the embryo or to the infant: after all, if we never informed them, it would thereby preserve their existential freedom through ignorance. This brings us to the second of Jonas's objections to genetic engineering: that it would corrupt the relation between the generations by becoming one of manipulator and manipulated. It creates, as he puts it, an entirely one-sided control 'of present men over future men, who are the defenceless objects of antecedent choices by the planners of today. [...] [P]ower is here entirely unilateral and of the few, with no recourse to countervailing power open to its patients' (*PE*: 147). Here the concern is not to do with self-understanding, but one's objective relation to others. In being manipulated in a particular way – again, even if not deterministically – the enhanced become the object of someone else's design. The worry is that this one-sided power relation once more undercuts freedom, although this time not of the existential sort. Rather, this state of affairs threatens our republican freedom of non-domination by others.

Now, the proponent of human enhancement could argue that we regularly engage in a practice that conforms to this type of power relation, one that we not only tolerate but actively champion: education. Just as in cognitive or moral enhancement, in education one person seeks to shape the character and intellect of another, so if the latter practice is permitted, presumably the former should be also. Making this very parallel, Ingmar Persson and Julian Savulescu argue that '[t]here is no reason to assume that moral bioenhancement to which children are exposed without their consent would restrict their freedom and responsibility more than the traditional moral education to which they are also exposed without their consent' (2012: 113). To challenge this equivalence I turn to Jürgen Habermas, who in *The Future of Human Nature* explicitly built on Jonas's insights.

Habermas observes that education and genetic engineering embody fundamentally different principles: the former operates according to the discursive principles of communicative rationality, while the latter conforms to the technical principles of instrumental rationality. What this means is as follows. Education, relying as it does on symbolic communication, presupposes the *mutual capacity for reason*. Even if the pupil does not at the time fully understand the reasons behind educational content, as a rational being in development they are in principle able to – and in time hopefully will – comprehend those reasons. When they do, they are then free to accept or reject those reasons. This shared ground allows for a fundamental equality between agents despite the asymmetry inherent in the practice of education. As Habermas says, '[E]xpectations underlying the parents' efforts at character building are essentially "contestable" [...] the adolescents in principle still have the opportunity to respond to and retroactively break away from it' (2003: 62). Indeed, one might go so far as to say that developing the ability to question what one has learnt – to think for oneself – *is* an objective of education, although this is not required for the present point. For even in instances where learning to think for oneself is not an objective of the educator, the nature of education itself means the content can nearly always be subsequently contested, as Habermas's own schooling in Nazi Germany starkly demonstrates.

Genetic engineering, by contrast, lacks the mutual ground of reason which allows for equality between participants. Instead, as a technical procedure carried out on the child as an object, the manipulation makes retrospective disagreement impossible:

> With genetic enhancement, there is no communicative scope for the projected child to be addressed as a second person and to be involved in a communication process. [...] It does not permit the adolescent looking back on the prenatal intervention to engage in a *revisionary* learning process. *Being at odds with* the genetically fixed intention of a third person is hopeless. (2003: 62)

The difference in principle, therefore, is that the power relation of genetic engineering is not only unilateral but also incontestable: the child is bound to an intent from which they cannot be released. There is, as Habermas says, no scope for autonomous contestation. By contrast, education, courtesy of its basis in rational communication, possesses an inherent reflexivity and thereby presupposes freedom as non-domination. Here Persson and Savulescu disagree. They claim 'common sense and science' tell us that 'it is surely evident that when small children are taught language, religion, basic moral rules, or whatever, this education is just as effective, irresistible, and irrevocable as biomedical intervention is likely to be' (2015: 52). But as Habermas shows, even education as fundamental as moral rules or metaphysical beliefs can always be questioned, one's relation to it being altered through the act of contestation.[7]

Genetic engineering, as a unilateral and incontestable intervention, therefore remains qualitatively distinct. But how are Jonas's concerns about existential and republican freedom connected to dignity? Clearly, if the intervention has any effect at all, then, in altering the organismic *telos* of the person, genetic engineering would represent a violation of non-personal dignity. But in particular cases it would violate human dignity also. Genetic engineering to be carried out eugenically – that is, to improve the species or national stock – entails being treated as a mere means to the ends of the group. This provides us with grounds to reject outright a state-sponsored programme of genetic engineering, regardless of whether the effects were heritable or not: the absence of free and informed consent, which alone can excuse such instrumentalization, would make it categorically impermissible. Less clear are the implications for human dignity in the event that genetic engineering is carried out for the sake of an individual's well-being, as parentally chosen genetic engineering would presumably be. (Obviously, if it were instead conducted with others' interests in mind, such as the parent's own, then it would be as impermissible as the state-sponsored programme.) We assume for the sake of argument that such well-meaning genetic engineering would be for a purpose that the individual is highly likely to retrospectively approve of. In that case the prospective children would simultaneously be treated as an end, with their likely interests motivating the intervention.

This leaves us with a human dignity-based objection to eugenic genetic engineering, but not to genetic engineering performed for the sake of the engineered person's well-being. And yet, we can still draw on the violation of non-personal dignity to account for our discomfort. In seeking to mould an individual through genetic engineering, even if for their benefit, we risk curtailing existential freedom and republican freedom in unprecedented ways. This would hold most obviously for any decision made by one person for another, as in the parent–child engineering scenario, but could hold even in the case of somatic genetic engineering carried out on oneself. Even though the individual would by necessity treat themselves as an end, they would still be inescapably bound to their prior intention. Again, this would not provide us with a categorical objection based on human dignity, but one that nevertheless goes some way to accounting for our lingering disapproval.

Having discussed Jonas's central objections to its preeminent method, we shall now turn to his objections to the *ends* of transhumanism. As stated, the transhumanist seeks to extend human capacities and faculties beyond their given range. Rather than analysing each type of possible enhancement in detail – moral, physical, cognitive and so on – Jonas questions the wisdom of enhancement as such.

> Whether we have the right to do it, whether we are qualified for that creative role, is the most serious question that can be posed to man finding himself suddenly in the possession of such fateful powers. Who will be the image-makers, by what standards, and on the basis of what knowledge? [...] These and similar questions, which demand an answer before we embark on a journey into the unknown, show most vividly how far our powers to act are pushing us beyond the terms of all former ethics. (*IR*: 21)

Note once again the reference to the image of Man: here referring not to humanity's true image but rather to an *arbitrary* image that is then made flesh. Should the creation of human beings according to such an image contravene the true idea of Man, then it would straightforwardly violate the imperative to realize the latter. But what if it did not? What if human beings worthy of the name continued to exist, albeit modified through biotechnological art?

The fear of eugenics lingers in the background here – not unreasonably, given that Jonas lived long enough to see the utopian hopes of early-twentieth-century eugenicists devolve into the Nazi nightmare. Today, however, few transhumanists advocate a eugenics programme.[8] Enhancement, they say, will be *liberal*: a matter of personal autonomy and reproductive freedom. Jonas's arguments have already shown this to be dubious – though not categorically impermissible – by examining its envisioned method. Let us continue our reconstruction of his critique, this time on the basis of the envisioned outcome of enhanced human beings.

The first problem one encounters in developing a critique of non-eugenic enhancement is that the improvement of human beings appears to also be an end of healthcare. With medicine we treat, cure, restore and otherwise try to ameliorate a variety of conditions, both physical and psychological. Drawing a qualitative and not merely quantitative distinction between healthcare and enhancement is therefore far more difficult than one might assume. We stated above that enhancement entails the extension of capacities and faculties beyond their given range, but this notion already admits of some historical and cultural relativity. After all (the transhumanist will object), we have for centuries enhanced lifespans and physical capacities through medicine and increased living standards: thus 'the given range' of human capacities is different for a contemporary European citizen and a medieval serf. If healthcare is a laudable practice, as we of course take it to be, and the consequences for our health and lifespan good, as we assume them to be, then enhancement may simply be the logical next step.

Perhaps, however, there is a misunderstanding here of the nature of healthcare, which when properly understood points towards the difference between it and enhancement. Jonas's explorations of the organismic basis of health, when combined with Gadamer's hermeneutic investigations into the topic, allow us to account for this difference.

The difference between healthcare and biotechnological enhancement is perhaps obscured by the fact that both would be, according to the Western canon, forms of *technē*. But there remains an essential difference between healthcare and all other arts and crafts, for in it nothing is produced. To be sure, its goal is health, but health is not a novel entity like an artwork or a machine. As Jonas says, '[H]ealing is not the production [*Herstellung*] of a thing, but the restoration [*Wiederherstellung*] of a state, and the state itself, although art is applied to it, is not artificial' (*TME*: 146). Since health is not produced by artifice, it follows that it is instead 'defined by nature', and more specifically the immanent *telos* of the organism: '[F]unctional integrity alone is its object' (*TME*: 147, 149). Jonas's philosophical biology and anthropology here allow us to draw out, to a significant degree, the difference between healthcare and other forms of *technē*. In healthcare we follow the *telos* of the organism, assisting it where possible in its orientation towards wholeness. Thus we set broken bones, stem the flow of blood from wounds, undergo psychiatric therapy, take medicines, and employ prosthetics, transplants and life support machines to allow our bodies to restore or maintain integral functioning for as long as we are alive. By contrast, enhancement uses the *telos* of the organism as a starting point, as a basis from which to create something new. For the transhumanist our capacities and faculties are not definitive of our being and the locus of medical attention, but rather limitations to be overcome.

This neat distinction is undermined, however, by the fact that healthcare already *does* go beyond our teleologically defined capacities, which Jonas even acknowledges (*TME*: 150). The classic example is acquiring immunity to a disease through vaccination, which arguably represents a novel capacity rather than the restoration or maintenance of a state. Is the transhumanist then correct to say that enhancement is simply a logical extension of healthcare?

Here Gadamer's complementary insights take us one step further. Like Jonas, Gadamer holds that healthcare is generically different from other forms of *technē* given its orientation towards the *telos* of the organism (1996: 32–3). But this, he says, is a legacy of our pre-modern – more specifically, Greek – medical heritage, which is today intertwined with a competing tradition that has its origin in the modern scientific revolution. We recall from Chapter 1 that Jonas identifies 1543 as a symbolic turning point in Western history, as the key texts of Copernicus and Vesalius were published that year. This represented 'the two sides of the scientific revolution as it eventually took shape: the macrocosmic and the microcosmic' (*PE*: 52). Jonas does not, however, fully sketch out the theoretical – as opposed to the practical – implications of this event for healthcare. Gadamer fills in this gap, noting that modern medicine takes on the character of modern technology: '[I]t understands itself precisely as a kind of knowledge that is guided by the idea of transforming nature into a human world, indeed almost of eliminating the natural dimension by means of rationally controlled projective "construction"' (1996: 39). The consequence of this understanding of human beings – just as in modern science's understanding of *non*-human beings – is that 'this knowledge allows us to calculate and control natural processes to such an extent that it finally becomes capable of *replacing* the natural by the artificial' (1996: 39).

The intertwining of these two conceptions of healthcare, the ancient and the modern, rooted as they are in different understandings of nature, give contemporary healthcare its ambiguous character. It 'can never be understood entirely as a technology', but instead 'represents a peculiar kind of practical science for which modern thought no longer possesses an adequate concept' (1996: 39). This has profound implications for our present concerns. While much medical practice is still consistent with the ancient conception of healthcare – even if it does not explicitly understand itself that way – enhancement can be understood as the logical extension of healthcare *in the modern sense alone*. In trying to overcome rather than aim at the *telos* of the organism, it belongs purely to medicine's modern 'productive' dimension, along with practices such as non-reconstructive cosmetic surgery.

If this tells us what the logical difference is between healthcare and enhancement, it still leaves us with the normative question of whether the latter is any less commendable than the former (assuming, of course, that we have already ruled out any form of enhancement that violates human dignity). Here there is no simple answer based on the different principles underlying ancient and modern medicine. To be sure, in working with the *telos* of the organism, ancient medicine respects our non-personal dignity, whereas a practice that can only be understood according to the logic of modern medicine violates it. This is, however, insufficient reason to reject the latter: after all, vaccinations also exclusively belong to the modern 'productive' understanding of healthcare, and we are clearly unwilling to condemn them for that or any other reason.

Why, then, are we specifically troubled by the transhumanist vision? The answer, imprecise though it may be, is that enhancement disregards our organismic *telos* and non-personal dignity *as such*. The transhumanist isolates one aspect of the human condition – our cognitive freedom – and declares the rest superfluous, to be cast aside like obsolescent technology. It is for this reason that it follows logically from modern gnosticism: it conceives of our 'real' selves as a thinking substance to be liberated from nature's corporeal prison. This is, indeed, the logical extreme of modern materialism, presupposing an entirely mathematically calculable, rather than teleological, conception of the organism that can be remade at will. While we are, of course, thankful for the vast majority of the advances that the modern conception of healthcare has delivered – particularly when these are guided by the Greek understanding – alone it is based on a deficient understanding of the human being. Our real selves are, in truth, embodied, vulnerable and dependent. As Gadamer shows, the Greeks knew this well, reflected in their experience of health as a state of equilibrium, and disease as a disruption of it (1996: 36). By contrast, having abandoned the Greek conception of healthcare entirely, the transhumanist experiences *their very organismic being as a disease*: every limitation is perceived as an imperfection, every vulnerability a cause for what Günther Anders once called modern humanity's 'Promethean shame' before our perfect machines (1956: 23).

Perhaps the most egregious example, and the aspect of enhancement about which Jonas wrote most extensively, is the desire for immortality. Jonas does not doubt the value of immortality understood in the Classical sense: that which is made possible by our public life in the *polis*. On the contrary, he devotes the concluding essay of *The Phenomenon of Life* to that very topic, celebrating 'the meteoric flash of deed and daring [...] immortalized by worldly fame' (*PL*: 271). Immortality understood in this sense is congruent with the finitude that characterizes life: '[N]ot what lasts longest in our experience, but what lasts shortest and is intrinsically most adverse to lastingness, may turn out to be that which binds the mortal to the immortal' (*PL*: 271). For humans, uniquely open to their own *eidos*, possess an understanding of their condition and act against this shared backdrop regardless of geographic, cultural or social circumstances: we know that we are born rather than made, that we will one day die and that every life shares this givenness. But the duration of time between the twin poles of natality and mortality represents a horizon of freedom, within which we are capable of deeds that can far outlast our mortal remains, granting us immortality should our acts become legend.

Transhumanists argue for immortality in a very different sense: as the permanent extension of the individual's life. Just like a machine, which as non-living can in principle persist indefinitely, the transhumanist seeks to become an eternally enduring post-human being, or even a mind contained within an invulnerable inorganic body. The intention, therefore, is to *overcome* the human condition. Advocates of this form of immortality tend to follow a simple utilitarian logic: if a long, healthy life is a good by virtue of the opportunities and experiences it can offer, it stands to reason that an indefinite continuation of this state must be even better. As Nick Bostrom puts it, 'Any death prior to the heat death of the universe is premature if your life is good' (2008: 4). The assumption here is that our pleasurable experiences would remain qualitatively the same (or even improve if we are cognitively enhanced), yet increase quantitatively. But this may, in fact, overlook the role played by our natality and mortality in giving meaning to those experiences.

We can take firstly the fact that we are born, and ask what significance our natality, that 'perennial spring', has for us (*IR*: 19). Jonas suspects that 'if we abolish death, we must abolish procreation as well, for the latter is life's answer to the former' (*IR*: 19). This is not just an ecological concern, to do with the mere fact of insufficient resources and living space on an already overcrowded planet. Although a pertinent objection, the transhumanist can always invoke a utopian solution in response: that if we have mastered death, we would surely be able to solve such logistical issues, probably by uploading ourselves to cyberspace. Jonas's worry is principally that the desire to procreate – as opposed to the desire for sex – is motivated, in part, by a concern for immortality in the Classical sense: to leave a mark on the world through one's descendants. The realization of transhumanist immortality may well diminish the desire for Classical immortality through procreation, and thereby result in a drastic reduction of births.

Now, the transhumanist may well see no problem with this, and argue that if human lives continue permanently, then it does not matter whether it is old or new. But Jonas argues us that it does matter, in terms of the constitution of society and the effects it would have on our culture: '[W]e would have a world of old age with no youth, and of known individuals with no surprises' (*IR*: 19). He continues:

> [Natality] grants us the eternally renewed promise of the freshness, immediacy, and eagerness of youth, together with the supply of otherness as such. There is no substitute for this in the greater accumulation of prolonged experience: it can never recapture the unique privilege of seeing the world for the first time and with new eyes; never relive the wonder which, according to Plato, is the beginning of philosophy. (*IR*: 19)

Conversely, were a wellspring of youth *and* immortal beings to exist concurrently, the result could be an ever-greater estrangement of the old from the young, the former 'stranded in a world we no longer understand' (*MM*: 98). Neither is a desirable prospect.

At the opposite end of our temporal existence is the pole of mortality and being-toward-death. What would the eradication of this limit mean for us?

In addition to the considerable consequences that biological immortality would doubtless have for society, Jonas's existential perspective shines a light on the effects it could have for our self-understanding. In an existentialist vein, in line with the memento mori tradition, Jonas argues that knowledge of our eventual deaths plays a fundamental role in giving meaning and weight to our lives (*IR*: 19). The reason is that our finitude is in fact a presupposition to our making meaningful decisions, which then inform who we are. To be sure, I cannot choose to have been an ancient Briton or a contemporary Amazonian, but to the extent that our lives are undetermined we have before us a range of possibilities: if we are lucky we can choose where to live, which job to take, whether to have children and so on. On a more everyday level we make choices ranging from how to treat others to which pastimes to engage in. One might choose to spend precious free time learning to play the piano rather than video games, to see family or friends rather than remain alone. The fact that this choice is delimited by the given duration of our lives is precisely what grants the decision weight, because we have chosen to allocate the cherished time we have thus.[9] An immortal, however, could eventually do anything and everything. And yet rather than this being liberating, as the transhumanists naively suppose, it could in fact sap actions of their meaning: if we cannot die, an infinity of options are open to us and thus no longer have the weight we presently experience in them *as choices*. What would be gained quantitatively would be diminished qualitatively, and so the abolition of mortality would amount to a form of existential denigration. For this reason, Jonas suggests that '[p]erhaps a nonnegotiable limit to our expected time is necessary for each of us as the incentive to number our days and make them count' (*IR*: 19).

Taken together, the foregoing critique of genetic engineering as a means and human enhancement as an end gives us ample reasons to reject the transhumanist project. If the goal is as dubious as Jonas suggests, and genetic engineering as fraught with risks to freedom as we suppose, then it would be wise to rule it out in principle. We must note, however, that the threats to existential and republican freedom are less compelling when divorced from the arbitrary purpose of enhancement. That is to say: although the freedom-based concerns apply *simpliciter*, they derive additional persuasive force from being considered in tandem with enhancement as the purpose of the intervention. But if genetic engineering were undertaken not to enhance, but rather to cure or prevent diseases, then the intervention would perhaps be perceived by the engineered as a tolerable restriction on freedom.[10] This 'therapeutic' kind of genetic engineering might be less an enhancement in the sense that we have discussed – motivated by a desire to transcend the human condition – and perhaps closer to a 'modest' enhancement such as acquiring an immunity through vaccination. Although vaccinations still conform to the logic of enhancement, their preventative rationale represents a less severe break with the human condition, and it is possible, at least, that therapeutic enhancement would be seen in the same light.

If, therefore, a form of genetic engineering could partially evade the aforementioned concerns, it would be the therapeutic sort. As Jonas says, it is certainly less objectionable in terms of its goal, and the utilitarian promise might be sufficient to override any lingering principled concerns about freedom (*EBA*: 503). But there remain worries to do with possible consequences that could yet lead to a staying of our hand. Complexity and unpredictability appear to define the genotype in its relation to the phenotype. As such, even therapeutic genetic engineering with the most precise tools would risk errors which cannot be undone – the defective human being can hardly be scrapped like a faulty machine – or which do not become apparent until long after standard practice had been established. This would be unforgivable if such mistakes were committed to the germ-line and thus hereditary, transmitted through the generations. It would, perhaps, be less foolish if only made at the somatic and therefore nonheritable level, under conditions of free and informed consent. In response to this prospect Jonas appeals to a slippery slope argument. What might well begin as an understandable attempt to cure and restore at the somatic level could open Pandora's Box, 'leaving behind the conservative spirit of genetic repairs and embarking on the path of creative arrogance. [...] It would probably be wiser to resist even the charitable temptation for once, in this instance' (*EBA*: 503-4).

Where, then, does this leave the idea of Man, the ultimate object of responsibility? In considering various techno-scientific threats to human dignity, what did we discover about the fullness of our true image? Jonas firstly shows us that our unique personal dignity categorically rules out the instrumentalization of human life in medical research without free and informed consent. For the same reason, it also prohibits genetic engineering for eugenic purposes. Moving from this solid ground, we also used Jonas's thought to argue – admittedly with caveats – that the echo or trace of human dignity discernible in the permanently non-conscious patient prohibits the harvesting of their organs, while permitting the withdrawal of their life support if the body is also ready to slip away. The idea of Man therefore demands that we never treat others only as means but also always as ends.

In addition, we followed Jonas in making a supplementary case on the grounds of non-personal dignity for abiding by the human condition. This is not, to repeat, a categorical argument, but one that nevertheless frequently has strong intuitive appeal. On this basis we justified Jonas's objections to genetic engineering that perverts the relations between the generations – which are, or should be, characterized by republican freedom – and undermines our existential freedom to forge oneself in ignorance of a preformed plan. We also argued that our finitude, so constitutive of the human condition, gave us good reason to protect natality and mortality against their abolition or radical transformation. In this way we sought to resist the eschatological promise of human enhancement, which rests, as we have seen, on a gnostic conception of the human being as a mind trapped in an obsolete body. The idea of Man and the human condition, however unquantifiable and conceptually imprecise they may be, therefore represent two significant guides for action in the bioethical domain – guides that will likely become only more necessary in the coming decades.

Conclusion

I Humanity: The shepherd of beings

We stated at the beginning of this book that the purpose of Jonas's philosophical enterprise was to counter modern gnosticism on the levels of ontology, axiology and ethics, the latter pertaining in particular to the ecological crisis and biotechnological revolution. Having since looked at Jonas's contributions to these fields we must now ask how successful his attempt is. It is not to be expected, of course, that Jonas answers any of the relevant philosophical questions in a way that holds for all time. Regardless of whether or not such answers even exist, one can never say, given the limitations imposed on thought by historical and cultural circumstances, that they have been arrived at beyond reasonable doubt. To do so would be wildly hubristic. But we can ask whether a philosophy is *good enough*, both by the general standards available to us and the particular aims it sets itself.

Let us begin our evaluation of Jonas's anti-gnostic system by firstly looking over its weakest aspects: namely, its politics and axiological objectivism. Starting with the former, we might ask whether it truly escapes modern gnosticism. Stephan Kampowski (2013: 112–13) and Gerald McKenny, for instance, both identify in Jonas's ethic of responsibility a Baconian will-to-mastery: according to McKenny, '[T]he responsible subject, […] defined by its care for what has come under the guilt of its power, is still essentially a modern subject who must gain control over technology' (1997: 74). This purportedly poses a problem since 'it determines Jonas' conception of politics. Political theory for him reduces to the question of which system, Marxism or capitalism, is more likely to gain control over the dynamics of technology' (1997: 74). McKenny suggests instead that 'the task is not to gain control over technology but to be guided by a process of moral formation that is capable of both resisting its diffuse power and assimilating it into a moral project' (1997: 74–5). According to this line of argument, then, Jonas only counters Baconianism with Baconianism in a different guise.

McKenny is not wrong to say that Jonas seeks, through a politics guided by responsibility, a second-order power over the technological power over nature.

To the extent that Jonas's one-sided focus on the positive freedom to act leads him to formulate an authoritarian theory of the state in *The Imperative*, it is indeed politically problematic. This consequence aside, however, McKenny's argument does not really amount to a criticism – on the contrary, political control of technological development is precisely what the present moment demands. Perhaps this does entail a conception of the Baconian subject exerting control over the human world, but that is not a problem for Jonas since he never claimed to escape instrumental rationality in each and every instance. His critique of Baconianism – and the modern gnosticism of which it is a key part – is restricted to specific domains: namely, a reductive conception of human and non-human nature and the fact that our desire to master these has led us to the brink of biotechnological transformation and ecological catastrophe. Jonas never claims that the solution will not involve *self*-mastery, and nor should he, since morality often demands exactly that.

That said, Jonas's political theory undoubtedly contains argumentative errors that lead to his acquiescence to authoritarianism, as we have seen. Perhaps most egregiously, in *The Imperative* Jonas seeks to justify paternalistic statesmanship with reference to responsibility for the infant. Yet political responsibility aligns not with this total and private form of the phenomenon but rather the public responsibility each moral agent has for one another by virtue of belonging to a group with power over the individual. Through the latter the common good is revealed to be the end of collective political action, rather than the aim of a statesman who perceives the truth and leads his people towards it. Why Jonas did not recognize the gnostic dimensions of his account of statesmanship is unclear. Regardless, he began to replace it with an intriguing environmental republicanism, albeit too late in life to fully develop the theory. This alternative does indeed align with an account of public responsibility, thereby representing a credible political ethic, and for this reason Jonas is perhaps not to be judged too harshly as a political theorist.

The second major criticism to be levelled at Jonas's system concerns his axiological objectivism. Jonas's argument from intuition has been extensively criticized by post-Kantian philosophers for its dogmatism, as discussed earlier, and in a slightly different form by Gilbert Hottois, whose objections we shall consider now. According to Hottois, Jonas's philosophy of responsibility ultimately relies on dogmatic Judeo-Christian foundations rather than comprising a fully secular and rational ethics. Hottois observes, rightly, that 'a careful reading of *The Imperative of Responsibility* shows that [...] religious stylistic overtones are not uncommon' (1993: 14). In particular he criticizes Jonas for holding that 'nature – here meaning living species – is sacred and inviolable' (1996: 136). Certainly, as we have seen, Jonas holds living beings to be ends-in-themselves and each to possess a good of its own – a fact which is then affirmed as objectively good – but he certainly does not regard life as sacred and inviolable. As discussed in Chapter 5, the integrity and non-personal dignity of living beings give us a reason to consider their good in any relevant moral situation, but this will not necessarily be a reason that overrides other concerns. Here there is no fear of playing God with a sacrosanct nature, but rather a legitimate concern with doing unnecessary or excessive harm to non-human life.

Hottois's broader charge – that Jonas's nominally secular ethics and axiology draws on Judeo-Christian notions – is undoubtedly correct to the extent that Jonas is influenced by religious sensibilities, both rhetorically and intellectually: a charge no one who has read his work could deny. But it is another point entirely to claim that this actually matters for the sake of evaluating his arguments, which stand or fall on their own merit. Irrespective of religious sanction one can offer, as we have suggested that Jonas does, a convincing account of the dignity and integrity of human and non-human life that provides adequate grounds for their being objects of moral responsibility. The fact that his arguments broadly track the Judeo-Christian tradition is not a problem for Jonas – or at least not for the interpretation of his thought offered here – because ethics *by its very nature* belongs to tradition, whether in affirmation or denial. In our case this is the Classical-Judeo-Christian West, and therefore it is no surprise that Jonas's ethics are so obviously indebted to this history. What matters philosophically is that the ideas are rationally defensible, and for the most part they are.

The most obvious exception is Jonas's attempt to provide an objective axiological foundation for his ethical theory. It is admirable, no doubt, but necessarily falls short. The is–ought gap may well be held as unquestionable by some contemporary philosophers, and advanced simplistically by others, but it endures because Hume's insight ultimately holds true. As our discussion of Alasdair MacIntyre in Chapter 4 showed, the attempts made by various modern philosophers to find an objective foundation for morality were doomed to fail because of a category mistake: axiology does not have its foundations in reason – although it is responsive to reasons – nor in a universal human 'moral sense', but rather sentiment and custom. As such, Jonas does not succeed in his ambition of overcoming modern nihilism, and nor, therefore, can his philosophical project be considered a complete success. Nevertheless, even if we can only accept Jonas's axiological axiom – that the existence of subjective value is objectively valuable – on *intuitive* grounds this does not, I think, fatally undermine his project. The intuition can always be denied, as Jonas says, but it is pervasive according to the tradition that constitutes our broader axiological understanding. We are not, then, left with subjectivism but rather relativism – more specifically, a 'deep' historico-cultural relativism – which provides a defensible axiological basis for an ethic of responsibility, albeit one that falls short of Jonas's lofty ambitions.

Even if Jonas's philosophy does not, then, fully escape the finitude that Heidegger's existentialism had prescribed as our lot, his philosophy of life and ethics nevertheless provides us with the resources to largely transcend the gnostic limits of Heidegger's thinking. Heidegger had taken us to be thrown into being with only the possibility of an authentic grasp of our being-toward-death as consolation. Jonas, by contrast, stresses that being human not only formally overlaps, to a considerable extent, with being a non-human form of life, but also that what we do *not* share with other living beings is itself the final achievement of a tendency in Being. That is to say: humans, as embodied selves, share much the same care-structure as other life forms, a structure that varies in richness throughout the domain of life and in doing so presents itself as an ascent towards greater world-openness. This, we may follow Jonas in supposing, is the result of a tendency in Being towards life and freedom. Hence the development of a symbolic existence, reaching a unique prominence and intensity in human life, and ultimately allowing for morality, is itself the *nisus* of Being at work. Even though the *content* of morality belongs to our historicity, our moral being itself is therefore the result of something greater than any one form of life.

The value of our moral being was then revealed in the case of responsibility for the newborn. Not only is this instance of responsibility the most fundamental ethical injunction prescribed by our moral tradition, but it also closely follows from our organismic being. And there is a further reason for its exceptional status. The incontrovertible nature of the phenomenon also indicates the supreme value of its true object: humanity as a moral being. For in each instance of responsibility for the newborn we are committed to not only this child before us right *now*, but also this child in its becoming a moral agent. Responsibility is therefore future-oriented, pointing towards the idea of Man as a moral being that humanity alone can fulfil. This was no mere species-relative good, but rather – as the outcome of its *nisus* – a good of Being itself.

Finally, we were able to account for the worth of non-human life and the human condition without resorting to species essences. Each individual organism, oriented towards its continued existence, is an end-in-itself and has a good of its own, while humanity possesses a unique personal dignity courtesy of its evolutionarily acquired moral being. All are for us a matter of global and intergenerational responsibility now that their bare existence is in doubt. On the one hand, the ecological crisis threatens the plenitude of life that constitutes the biosphere, and through this, the future of humanity; on the other hand, humanity might undergo a biotechnological degradation of its own doing. In both cases what is imperilled demands of us protection instead. These newfound responsibilities are proof, if it were needed, that humanity is not only the shepherd of being, as Heidegger had it, but also the being that is uniquely capable of moral responsibility: the shepherd of *beings*. In this task of guardianship lies the pinnacle of our transcendence.

II Carrying the fire

The previous sentence might have provided a satisfying note to end on. But it would not, in truth, have been an appropriate one. For Jonas's mature thought aims to not only show that we are the shepherd of beings, but it also implores us to act accordingly. It is concerned, perhaps above all else, with the threat to human and non-human life posed by the ecological crisis – a threat that has steadily increased in intensity in the years since his passing.

In Chapter 1 we briefly outlined various climate change forecasts based on possible pathways of reducing, or failing to reduce, our CO_2 emissions. The scenarios ranged from – at best – the considerable risk of extreme weather events, and harms to unique and threatened natural systems, to, at worst, a barely habitable earth. These predictions, based on scientific data, are sobering enough. But Jonas reminds us that data alone are insufficient: as emotional, perceptual beings we need *stories* to bring the scenarios to life. This is the rationale for his heuristic of fear, discussed in Chapter 6, which invites us to imagine scientifically credible outcomes of our techno-industrial activity so that we might act to prevent the worst of these from becoming a reality. Jonas mentions Aldous Huxley's *Brave New World*, which envisions a society dehumanized by the pursuit of utility through pharmaceuticals and biotechnology, but does not cite an equivalent work of fiction pertaining to the ecological crisis. One that crystalizes both his philosophy and fears for the future can be found, however, in Cormac McCarthy's *The Road*.

To those readers familiar with McCarthy's oeuvre and the extensive scholarship on it this claim may sound odd, as his novels, and the worlds they portray, are frequently described as gnostic. While this characterization may be true of McCarthy's earlier works, in particular the elementally violent *Blood Meridian*, the cosmology of *The Road* is of an altogether different – indeed, a distinctly Jonasian – variety. As such, a reading of the novel's key themes will draw together all that has been discussed so far with a view to the overriding concern of Jonas's final years, thereby bringing this work to a fitting close.

The Road is best described as post-apocalyptic fiction, although it stands out from the majority of such works for a variety of reasons. For one thing, the precise cause of catastrophe – whether nuclear war, climate change or even the former precipitated by the latter – is unknown to the reader. This matters little for our purposes, however, as the novel's setting is consistent with the worst-case scenario that runaway climate change could bring about. What is that scenario? McCarthy reminds us that it is not, in fact, civilizational collapse. For the downfall of civilization alone implies that nature remains intact, and with this the possibility of some kind of organized human society, a form of collective life, however rudimentary, that might nurture the good. The worst-case scenario is instead wholesale ecosystemic collapse. The latter entails not just that life as we know it comes to an end, but that *life as such* comes to an end, taking the opportunity for continued human existence with it. In this total apocalypse lies the horror of McCarthy's vision, given perfect expression in his mercilessly brittle prose.

Making their way along the road to the coast are the Father and the Boy. Everywhere they pass through they encounter only death and ash enveloping the fading memory of a world:

> The road crossed a dried slough where pipes of ice stood out of the frozen mud like formations in a cave. The remains of an old fire by the road. A dead swamp. Dead trees standing out of the gray water trailing gray and relic hagmoss. The silky spills of ash against the curbing. […] The ponderous counterspectacle of things ceasing to be. (2006: 293)

Such conditions present only two options for survival: cannibalism, as practised by the 'bad guys' stalking the Father and the Boy, or a relentless search for dwindling scraps of food. For those who choose the latter, life turns on chance discoveries: clean water, a handful of seeds, a withered apple – fragments of the past shorn of their meaning. There is no possibility of the self-renewal of nature, therefore no possibility of cultivating the earth and so no possibility of the continuation of human life. What little life remains is destined to slip towards extinction.

This wasteland without finds its reflection in a wasteland within. For the Father, our protagonist, being has reduced to the moment-to-moment concern for survival: 'No list of things to be done. The day providential itself. The hour.

There is no later. This is later' (2006: 56). Only when his immediate concerns are satisfied through a fire and a meal, with the Boy peacefully asleep, can the Father transcend mere animal existence and resume his fully human capacity for thought. But in such circumstances the mind finds no object of reflection save nothingness itself:

> He walked out in the gray light and stood and saw for a brief moment the absolute truth of the world. The cold relentless circling of the intestate earth. Darkness implacable. The blind dogs of the sun in their running. The crushing black vacuum of the universe. And somewhere two hunted animals trembling like ground-foxes in their cover. Borrowed time and borrowed world and borrowed eyes with which to sorrow it. (2006: 138)

Here depicted in *The Road* is precisely the new 'stage of primitivism' that Jonas fears might be our future: one of 'mass poverty, mass death and mass murder, the loss of all treasures that spirit has produced' (*CBE*: 22).

In spite of this objective hopelessness the Father persists, doggedly. Why? For the good, which shines all the brighter in the dark. The good does not transcend the world, like the personal God of Judeo-Christianity, who remains silent despite the Father's entreaties. 'He raised his face to the paling day. Are you there? he whispered. Will I see you at last? Have you a neck by which to throttle you? Have you a heart? Damn you eternally have you a soul? Oh God, he whispered. Oh God' (2006: 10).

In McCarthy's cosmology, on the contrary, the good belongs *to* the world, since it is, as in Jonas's philosophy, co-extensive with *life*, and above all with humanity. On the few occasions where non-human life appears in *The Road* – the dog, the trout of the novel's coda – these beings conjure a freedom and beauty otherwise absent from the world encountered. Most notable, perhaps, are the mushrooms that the Father and Boy discover, as these alone appear to be *surviving*, struggling yet succeeding, for a time at least, to fulfil their metabolic pursuit of continued existence. Of the area that provides for this life the Boy declares simply that it 'is a good place' (2006: 41). The connection between goodness and non-human life is reaffirmed and deepened by McCarthy once more in the text, so briefly that the reader might miss the clue altogether.[1] Almost in passing McCarthy describes the slow death of the biosphere as 'salitter drying from the earth' (2006: 279). But what is this *Salitter*? The word appears to have been used exclusively by the philosopher and mystic Jacob Böhme in his 1612 work *Mörgenröte im Aufgang*.[2] There *Salitter* refers to 'an all-encompassing divine substance' that takes the form of 'a matrix of forces that generate life and awareness' (Principe and Weeks 1989: 54). It is, in other words, Böhme's name for the god at work in the world, manifesting as living and sentient nature – a theological version of Jonas's *nisus* of Being. Thus whatever perceptibly exists of the divine is known to us through life, and sentient life in particular, and it is this, and the goodness it represents, that we have destroyed in McCarthy's vision of the future.[3]

As in Jonas's philosophy, however, it is humanity, whose freedom of thought represents the highest manifestation of sentience, that is uniquely valuable. And like Jonas, McCarthy identifies humanity's worth in its capacity for morality. While the majority of humans who appear in *The Road* fail to live up to that capacity, in the most horrific ways imaginable, the Boy remains pure of heart despite everything, and in doing so upholds the idea of Man: 'There were times when he sat watching the boy sleep that he would begin to sob uncontrollably but it wasnt about death. He wasnt sure what it was about but he thought it was about beauty or goodness. Things that he'd no longer any way to think about at all' (2006: 138).

Once more, for McCarthy this goodness can only be expressed in religious language. The Boy is described as the 'word of god', 'glowing in that waste like a tabernacle' (2006: 3, 293). What here appear to be references to the *imago Dei* are, however, inverted by a later passage – as though it is not humanity that is made in God's image, but rather that all deities are human abstractions from a holiness that is immanent in the world, and above all in ourselves. The Father 'watched him come though the grass and kneel with the cup of water he'd fetched. There was light all about him. [...] He took the cup and moved away and when he moved the light moved with him. [...] Look around you, [the Father] said. There is no prophet in the earth's long chronicle who's not honored here today' (2006: 297). It is accordingly the Boy, embodying the fulfilled idea of Man, who gives the Father his 'warrant' to keep going (2006: 3). The two of them are, as they tell themselves, 'carrying the fire': the fire of humanity, which must not be extinguished (2006: 87).

Thus *The Road* vividly reminds us, in line with Jonas's heuristic of fear, of what has utmost value and must never be placed at risk in our actions: the existence of humanity and life on earth. Yet we are at present imperilling precisely that, with an unsustainable form of life that we know we must renounce and yet show little collective commitment to doing so. We cannot wait, despite Heidegger's suggestion to the contrary, for the arrival of a god that might save us. As Jonas observed in his final public address, we will find the strength to change course, and so ensure a future for life on earth, only by heeding the call of responsibility, and thereby fulfilling the idea of Man:

> It was once religion which told us that we are all sinners, because of original sin. It is now the ecology of our planet which pronounces us all to be sinners because of the excessive exploits of human inventiveness. It was once religion which threatened us with a last judgment at the end of days. It is now our tortured planet which predicts the arrival of such a day without any heavenly intervention. The latest revelation – from no Mount Sinai, from no Mount of the Sermon, from no Bo (tree of Buddha) – is the outcry of mute things themselves that we must heed by curbing our powers over creation, lest we perish together on a wasteland of what was creation. (*MM*: 201–2)

Notes

Preface

1 In addition to the texts referred to in this volume, the finest English-language secondary literature on Jonas includes the following: Hauskeller (2015); Johnson (2014); Schütze (1995); Tirosh-Samuelson and Wiese (2008).

Introduction

1 Please note that any italics in quoted text are in the original unless otherwise indicated.
2 In making this claim I knowingly run against Jonas's own understanding of his work. He claimed that 'Aristotle didn't play much of a role' in his thinking, and was bemused that so many people should feel otherwise: '[T]here was little I could do to keep myself from being classified as a neo-Aristotelian. I wouldn't have classified myself that way, but it's hard to defend yourself against others' views. At any rate, I wasn't in bad company' (*M*: 204).
3 Although the seminar in question was nominally on Aristotle's *De Anima*, according to Kisiel (1993: 230–2) it ended up focusing on Book VII of the *Metaphysics*. In addition to this course, which took place in Freiburg, summer 1921, Jonas seems to have attended the following courses of Heidegger's in Marburg: the summer 1924 course published as *Basic Concepts of Aristotelian Philosophy* (2009); the winter 1924–5 course published as *Plato's Sophist* (1997b); the summer 1925 course published as *History of the Concept of Time: Prolegomena* (1985); and the winter 1925–6 course published as *Logic: The Question of Truth* (2010c). Jonas's lecture notes from some of these courses are held at the archive of his work in Konstanz.
4 Moreover, Jonas contributed one of his finest essays, 'Wandel und Bestand' ('Change and Permanence: On the Possibility of Understanding History' (*PE*: 240–63)) to the *Festschrift* organized for Heidegger's eightieth birthday (Klostermann 1970). Incidentally, according to Friedrich-Wilhelm von Hermann the essay particularly pleased Heidegger (2016: 93).
5 Heidegger there tied his history of Western metaphysics – according to which our openness to the manifold guises of being has devolved into mere technological appropriation of beings – to the 'emphatically calculative giftedness [of] the Jews' (2016: 44).

6 Jonas's use of the word 'nature' (with or without the capitalized 'n') seems to be coextensive with his use of 'biosphere', 'biosystem', 'the kingdom of life' and so on, rather than referring to non-living nature as well. Similarly, his references to 'Being' (with or without the capitalized 'b') appear coextensive with his references to 'Life' and 'life as a whole'. As a result, in Jonas's philosophy Being, nature, life and the biosphere typically coincide. It should be noted that his use of the word 'Being' is, in Heidegger's sense, metaphysical, since it refers to the totality of beings and so fails to observe the ontological difference between being (*Sein*) and beings (*Seiendes*). For this reason I capitalize Jonas's 'Being' except when he uses it in an existential sense (e.g. 'organismic being'), which I always write uncapitalized.

7 In reconstructing the key stages of Jonas's philosophy in this order, I take inspiration not only from Jonas's own retrospective assessments of his life's work (*WPE*; *PE*: xii–xx) but also from two earlier monographs by David Levy (2002) and Theresa Morris (2013). The latter both helped me to clarify the logic of Jonas's system and opened up pathways of interpretation that in the present volume I hope to have explored further, albeit in my own way. As such, I should like to acknowledge my indebtedness to Levy, for his discussion of Jonas's philosophical anthropology (2002: 62–77), and to Morris, particularly for her analyses of Jonas's teleological understanding of life, environmentalism and objectivist axiology (2013: 67–76, 96–118). Even though, as indicated, my approach to these issues often differs from theirs, I hope to build on them as part of a collective Anglophone effort to better understand, and so appreciate, Jonas's philosophical legacy.

Chapter 1

1 Jonas's doctoral thesis, jointly supervised by Heidegger and Rudolf Bultmann, was later published in two German-language volumes and the single English-language volume *The Gnostic Religion* (*GR*), which eschews much of the Heideggerian terminology.

2 Naturally, some philosophers of the Hellenic and Hellenistic periods partially departed from this outlook. Most famously Diogenes the Cynic argued that civilization had corrupted human beings, who would do better to live like dogs, free from the artificial restrictions of custom. Even here, however, Diogenes associates the good life with nature, which starkly contrasts with the Gnostics' wholesale repudiation of that connection.

3 Of course, Heidegger famously insisted that he was not an existentialist (1977: 151) and Jonas acknowledges that the term was not Heidegger's own (*MM*: 46). With some leniency, however, we can grant Jonas the usage of the phrase 'Heidegger's

existentialism', which he accepts can only be reasonably applied to Heidegger's early work (*PL*: 231).

4   My understanding of Heidegger's early work – up to and including *Being and Time* – is heavily influenced by Hubert Dreyfus's classic commentary, *Being-in-the-World* (1991), which to my mind remains the touchstone for clear and readable Heidegger scholarship.

5   On this basis Elad Lapidot (2017) argues that the purpose of Jonas's philosophical project is to confront and overcome the Gnostic principle per se, but this reading is too broad. Jonas's underlying aim, for reasons that will shortly become apparent, is instead to confront the Gnostic principle in its modern manifestation alone.

6   Jonas might be reasonably mistaken for a techno-dystopian given that in his best-known essay on the topic, he states: 'If Napoleon once said, "Politics is destiny", we may as well say today, "Technology is destiny"' (*TPT*: 35). However, Jonas did not mean this literally, a point he made explicit by correcting himself in a later essay. Referring back to the earlier claim he says he was 'speaking figuratively and exaggerating somewhat' (*EBA*: 491).

7   Because of its comprehensiveness Baconian techno-utopianism conforms to what Karl Mannheim, Jonas's postdoctoral supervisor, called the 'total conception of ideology […] the ideology of an age' ((1936) 1991: 49). Although Jonas studied under Mannheim and wrote one of his earliest articles on Mannheim's work (*KM*), the latter's influence on Jonas is not remotely comparable to that of Heidegger or Aristotle. Nevertheless, the connection is worth mentioning and – with the exception of Wennemann (2013: 114–22) – it has largely escaped the notice of commentators. It must be noted, however, that while Mannheim's conception of ideology sheds light on Jonas's social thought, their theories of utopianism are very much opposed.

8   The term 'genetic engineering', widely used for decades, has in recent years been supplanted by 'genome editing'. However, this does not so much reflect a qualitative change in the technology as an attempt at positive rebranding by industry ('editing' carrying less radical connotations than 'engineering'). For this reason I deliberately employ the former label.

Chapter 2

1   Both *Organism and Freedom* and *The Phenomenon of Life* have their origins in the *Lehrbrief* written by Jonas to his wife Lore during the Second World War. The former text is an integrated presentation of ideas developed in a series of essays written in the 1950s, but Jonas failed to find a sympathetic publisher for it: the text has only recently been published in the form of an appendix to volume I, part 1 of

his *Kritische Gesamtausgabe*. Jonas subsequently opted to publish the individual essays, accompanied by a few others, as a single volume titled *The Phenomenon of Life*. For a detailed account of the origins of the latter, see Franzini Tibaldeo (2009). Note also that the German translation of *The Phenomenon of Life* originally bore the title *Organismus and Freiheit* (later changed to *Das Prinzip Leben*), which should not be confused with *Organism and Freedom*.

2 After Heidegger and Aristotle the chief inspirations for Jonas's philosophy of life are, to my mind, the following: the biologist Ludwig von Bertalanffy, particularly regarding metabolism (see, for instance, the latter's *Problems of Life* (1952)) and biological systems (see Jonas's article 'Comment on General System Theory' (*CG*)); the zoologist Adolf Portmann, specifically on the nature of animal life (see, for example, his *Animals as Social Beings* (1961), and correspondence with Jonas regarding the *Organism and Freedom* manuscript (1956)); and the philosophical anthropologist Ernst Cassirer, whose affinity with Jonas on the topic of humanity is mentioned in Chapter 3. It is no coincidence that all were native German speakers, as Jonas's philosophy of life is clearly more in tune with the Germanic approach to biology, following Goethe, than it is the Anglophone tradition. Indeed, Jonas cites Goethe as a formative influence on his philosophical outlook and interests (*M*: 207–8).

3 It should be noted that Heidegger did broach the phenomenon of the lived body (*Leib*) twice in his lectures. The first was in his summer 1924 lecture course on Aristotle (2009: 136–40), and the second was in the *Zollikon Seminars* of 1959–69. In the latter he stated that *Dasein*'s being-in-the-world is 'determined by the bodying forth [*Leiben*] of the body', which was indeed a radical rethinking of the existential phenomenology of decades before (2001: 91). However, the connection Heidegger there drew between existence and embodiment was brief; in particular, he still had little to say about how *Dasein* and the lived body, if interlinked, related to the objective body (*Körper*). At no point, therefore, including in the *Zollikon Seminars*, did Heidegger undertake a full investigation of the body's significance, which remained for him 'the most difficult problem' (Heidegger and Fink 1993: 146).

4 Although a full discussion would represent too great a digression, it is worth noting that Jonas has been highly influential on the concept of *autopoiesis* and the enactivist theory of mind. See the following for both favourable and critical references to Jonas in these contexts: Weber and Varela (2002), Thompson (2004), Paolo (2006), Jesus (2016), and Villalobos and Ward (2016).

5 Here parallels with the Romantic philosophers of nature – particularly Schelling – are unmistakable. Aside from occasional references to Goethe, however, the Romantics do not, to my knowledge, explicitly feature in Jonas's work.

6 This is why Renaud Barbaras is wrong to suggest that Jonas misses 'life's essential mobility' (2008: 12). Metabolism is a processual movement through time and space,

and – crucially – as teleological can be distinguished from the movement of non-living beings, as we shall see.
7. For example, Joseph Farrell has argued that Jonas's account of the organism's self-organization should be explained with reference to the genome (2015: 191). But this explanation is insufficient, for the reasons given in this chapter.
8. This is, incidentally, why animists are wrong to attribute interiority to non-living nature.
9. Perhaps the closest exceptions are the dormant state of a seed, during which it still respires, and the capacity of microorganisms to conduct minimal metabolic activity when frozen.
10. Jonas does use the word '*nisus*' at one point, but does so, rather inconveniently, to refer to what I call the immanent teleology of the organism (*IHJ*: 354).

Chapter 3

1. Jonas does not discuss archaea, bacteria, fungi and algae, which are all now recognized as generic groups alongside plants, animals and humans. Presumably this is because he saw archaea and bacteria as no more complex than the organism per se, and fungi and algae as existentially identical to plant life. For this reason, the reader is invited to think of the section on plant life – and all references to plants – as referring to vegetative life broadly construed.
2. At only one point in his published writing does Jonas point towards it, describing animals 'who are able to play, namely animals with brood rearing, especially mammals with their sheltered childhood, who are still free from the grim pressure of animal needs but enjoy already the powers of movement' (*PE*: 249). (In the unpublished manuscript 'The Female of the Species' (*FOS*) Jonas mentions that this insight came to him while watching dolphins play from the deck of the ship on which he emigrated from Germany to North America.) Here is a partial recognition of the social aspect of animal being, yet the conceptual possibility is not fully pursued.
3. A transcript of the debate is included as an appendix to the fourth and fifth editions of Heidegger's *Kant and the Problem of Metaphysics*. Quotations here are from the fifth edition (1997a).
4. The earliest evidence of behavioural modernity, such as the artistic and decorative phenomena discovered in the Blombos Cave in South Africa, dates to the Middle Palaeolithic age. Typically, however, behavioural modernity is dated to the beginning of the Upper Palaeolithic age c.40,000 years ago. Whether and to what extent *Homo neanderthalensis* – better known as the Neanderthal – shared in the practices constituting behavioural modernity is currently a matter of intense debate, hence no position on the matter is taken here.

5 Note that Jonas's focus on representation distinguishes his anthropology from that of Hegel, who held that symbolic art per se is the earliest concrete manifestation of the Idea. See the first volume of Hegel's magnificent *Aesthetics: Lectures on Fine Art* (1975).
6 The distinction between 'art' and 'image' is worth noting here. As is well known, modern art gradually eschewed representation as a criterion, so that although paintings such as Mondrian's are works of *art*, they are not *images*, since the latter entails likeness. Conversely, paint daubed on a canvas by an elephant or chimpanzee is not an image – though whether it can be counted as art is another, far more complex, matter.
7 Of course, some individual chimpanzees and gorillas have been taught sign language, indicating beyond reasonable doubt that a capacity for symbolic thought and expression is cognitively possible for at least some animal life (Savage-Rumbaugh et al. 1998). But what is peculiarly human is the *prominence*, even ubiquity, of this supra-sensible medium, and the fact that it is inherent to our form of life rather than the product of artificial experimental situations.
8 Here Jonas diverges from Heidegger's conception of truth as unconcealment, as Lindberg notes (2005: 177), opting instead to defend its traditional form.

Chapter 4

1 Jonas's own example of organic self-organization is the contribution of the digestive system, which, oddly, he refers to as 'the digestive organ', as though it were one such entity rather than a system of organs (*IR*: 65). Presumably this is a misleadingly literal translation of the original German compound noun '*Verdauungsorgan*' (*PV*: 130).
2 For similar remarks, see Jonas's interview with *Der Spiegel* (*CBE*: 22–3), and his lecture 'On Suffering' (*OS*: 24, 30).
3 Vittorio Hösle has defended this aspect of Jonas's ethics from the position of discourse ethics, arguing that what Jonas has identified here is a transcendental limit of reason. Hösle suggests that Jonas's axiom 'could be grounded with the transcendental-pragmatic reflection that arguments themselves have a teleological structure which is already presupposed when we try to deny it' (2001: 44). Jonas does not pursue this line, however, presumably because he is unashamedly wedded to a fusion of ontology and axiology.

Chapter 5

1 Of course, ancient and medieval ethical systems, as well as at least some schools of Buddhist thought, stressed self-development through the cultivation of virtue. *Aretē* is, however, quite distinct from *technē*, which refers to the productive arts and crafts.

2 Plato is sometimes cited as providing an early awareness of anthropogenic environmental degradation, but his lamentation of the state of Attica's forests and topsoil in the *Critias* is not really to do with human influence, but rather natural disasters (1997: 111a–12d).
3 Jonas does not consider the possibility that procreation might be accounted for by virtue ethics: as a widely accepted component of the good life each of us has a strong reason to have children, who in turn would have reason to have children, and so on. Since this does not amount to an *obligation*, however, we cannot say that the future existence of human beings is thereby guaranteed, which is the very thing Jonas seeks in his new ethic.
4 My reference here to 'upholding' a right is deliberately left ambiguous, as depending on the nature of the right – for instance, a claim right or a liberty right – it can entail different things. A liberty right (such as freedom of speech) merely entails that others do not actively prevent someone from enjoying that freedom, whereas a claim right (such as a right to legal representation in court) entails that others are obliged to assist them in achieving it.
5 Furthermore, since rights arguably rest on a relationship of reciprocity – your right entails my duty and vice versa – by virtue of their inexistence future generations are incapable of having duties towards us, which would again mean that they are incapable of bearing rights. Now, the principle of reciprocity is not a universally accepted component of rights theory, as becomes clear when considering the rights we attribute to infants, the severely mentally ill and animals. As it happens, Jonas does subscribe to the principle of reciprocity (*IR*: 38), arguing that it applies to infants as they are *potentially* able to have duties towards us. However, he suggests that as animals are incapable of ever having duties, they therefore cannot have rights either (*OS*: 26–7).
6 Christine Korsgaard has pointed out that Kant's imperatives are not, in fact, synonymous and that the first is more permissive than the others (1986: 327). However, Jonas's criticism is centred on that formulation, and therefore it also applies to the stricter versions.
7 Jonas only briefly touches upon the metaphysical question of moral responsibility in *The Imperative of Responsibility*, and appears to adopt a compatibilist position: '[F]reedom […] is not absolute but confined to the latitude which physical necessity itself allows it' (*IR*: 221). His most extensive discussion of the issue is contained in the lecture course *Problems of Freedom*, but there his own position emerges only obliquely via interpretations of Aristotle and the Stoics. A full treatment of this issue represents too great a digression here, however, and so we shall proceed on the assumption that Jonas's compatibilist claims are tenable on the basis of his integral-monistic ontology, which already portrayed life as an ascending scale of freedom.
8 Clearly, this also means that any non-human life – or even artificial intelligence – that had the capacity for morality would also count as an embodiment of the idea

of Man. It has been argued that some non-human animals may in fact be moral agents, insofar as their behaviour is logically consistent with acting for moral reasons (Rowlands 2012). I am dubious about this claim, but if correct, it would certainly revolutionize Jonas's ethics. More troublingly, any humans who lacked the capacity for moral agency would not count as candidates for the idea of Man. As I have indicated, however, our responsibility in these cases remains intact on legitimate subjective grounds, even if it cannot be rationally accounted for by Jonas's theory.

9   I am grateful to Roberto Franzini Tibaldeo for originally bringing this connection to my attention, although he and I draw different conclusions from it. For Franzini Tibaldeo Jonas's conception of the image of Man is dynamic and subject to revision, whereas I understand Jonas as saying that this process of refinement allows, at least in principle, for the discovery of the *true* image of Man – namely, humanity as responsible. I should also note that Morris interprets this connection in the same manner as Franzini Tibaldeo (2013: 178–82).

10 As with Kant's categorical imperatives, Jonas's are not synonymous: the third version refers to 'humanity' per se rather than 'genuine human life' or the 'wholeness of man', which are intended to capture the fact that our responsibility is to humanity as a moral being (*IR*: 11). Moreover, only the third makes the requirement that this has to be on earth explicit (although the first also does in the original German, hence why I add this qualification in brackets to the quotation in the text).

11 Jonas does suggest, however, that poorer nations are obliged to assist in the fight against climate change in at least one respect, namely, a reduction of extremely high birth rates (*IHJ*: 366).

12 '[Think of w]hat we do to our domestic animals – the way we deprive our chickens of any life of their own by not having them in the barnyard anymore[,] and not in the chicken coop[,] but having […] egg factories in which a chicken never even experiences a life of a chicken in the open world. [Think of o]ur meat factories. Or the denial of sexual intercourse, the fruits of which we constantly demand from [animals], but with artificial insemination. I recently visited such an establishment for horses, and learned that the females never meet a male and the males never meet a female. There we do something on a grand scale […]. I have other impressions on a visit to Europe where I can see how the cows are still grazing on meadows, how they are driven out to the mountain meadows in the early summer, and have a life outside and are returned to their stables and so on, and it's a joy to see those cows' (*OS*: 25).

Chapter 6

1   I shall nevertheless remain closer to Jonas's thought than Murray Bookchin (1982) and David Levy (1987), both of whom use aspects of his thinking to develop

their own quite different political philosophies: social ecology and conservative environmentalism, respectively.
2 Exceptions include an obituary to Arendt (*HA*) and reflections on her more overtly philosophical thought (*AKT*). The two also briefly appeared in public debate together (Arendt 1979). Jonas's references to Strauss are even less frequent.
3 The original German for the precautionary principle is '*Vorsorgeprinzip*'. Nowhere, to my knowledge, does Jonas use this precise formulation, but his great work is of course *Das Prinzip Verantwortung* (literally, the responsibility principle), in which he repeatedly refers to *Vorsorge* (*PV*: 85, 90, 218, 219). In the English translation Jonas and Herr slightly awkwardly render *Vorsorge* as 'farsighted providence' and 'promotional care' (*IR*: 39, 121).
4 Both Jan Schmidt (2014) and Nathalie Frogneux (2014), from the German and French philosophical traditions, respectively, insightfully discuss Jonas's thinking in this regard.
5 Although Jonas uses the terms 'Marxism', 'socialism' and 'communism' interchangeably, by all of them he really means the ideology and political economy of the USSR, which I here call Marxist-Leninism or Soviet communism.
6 Indeed, the very title of *Das Prinzip Verantwortung* is a riposte to Bloch's *Das Prinzip Hoffnung*.
7 It might be objected that in taking Bloch as his representative of Marxist thinking Jonas unfairly represents the school of thought. There are, after all, those such as André Gorz who have sought to reconcile Marxism with environmentalism. Gorz argues that '*the ecological movement is not an end in itself but a stage in a larger struggle*', and that only a 'cultural revolution that abolishes the constraints of capitalism' could establish 'a new relationship between the individual and society and between people and nature' (1980: 3–4). While Gorz's socialism eschews any Baconian influence in terms of production, it nevertheless remains explicitly utopian. In an extended narrative section Gorz describes in detail 'one of several possible utopias', including a twenty-four hour working week, free public transportation, decentralized and self-sufficient economic units, organic farming, lifelong holistic education and so on (1980: 42). How this is to be achieved is not clear, but as an ideal it arguably escapes Jonas's critique.
8 Perhaps here, too, is a trace of Heidegger's influence on Jonas in the former's claim that both the USSR and the United States were 'metaphysically the same' (2000: 48) since 'determined by planetary technology' (1990: 54).
9 To his credit, Jonas also warns us that 'we should be careful not in our jubilation [following the USSR's demise] to think that it is capitalism […] which we should now salute as having been vindicated' (*CR*: 217).
10 Jonas's main example of the latter is Churchill (*IR*: 97), who was a 'hero' of his (*M*: 173).

11 Jonas notes that any description of a political community as youthful, mature or decrepit is merely figurative, there being – contra Hegel and Marx – no *telos* to history (*IR*: 109–11). He argues that where there appears to have been such, as with Lenin's action in October 1917 apparently fulfilling the prophecy of a proletarian revolution, this is only a post hoc judgement. The fact that Lenin acted when he did by itself *made* the theory correct, as the prediction, provided it holds enough appeal, 'acquires causal power itself in order to help its truth to gain reality, thus with intent contributes to the coming true of its prognoses' (*IR*: 115).
12 This interpretation can also be found – albeit in an undeveloped form – in Kurasawa, who classifies Jonas as a '[l]eft dystopian' (2007: 113).
13 Jonas also suggests that we might witness the arrival of a 'bizarre new religion' demanding 'the utmost in asceticism' – a new manifestation of the gnostic principle – but dismisses this option out of hand: '[T]here's no point in speculating about such things' (*CBE*: 23–4).
14 The 1983 essay in which he argues this – 'Auf der Schwelle der Zukunft' – represents, in my opinion, the turning point in Jonas's political thinking.

Chapter 7

1 One might note that Jonas's arguments have influenced the American 'bioconservative' school of thought, most notably through the work of Francis Fukuyama (2002) and Leon Kass. Indeed, Kass co-dedicated *Life, Liberty and the Defense of Dignity* to Jonas for his 'moral passion and philosophical courage' (2002: 299).
2 The most obvious example of such a threat would be a global pandemic of a deadly disease manufactured in the lab.
3 See, for example, John Harris's 'The Survival Lottery', which could easily be mistaken for a parody of utilitarian thinking (1975).
4 Note the requirement that the polity must itself be *just*: conscription is not a moral right we would be willing to extend to tyrannical regimes (although they would probably be the most likely to make a claim to the *legal* right), since the military defeat of such a state might actually benefit its citizens and protect human dignity. The crucial point, therefore, is that it is not the needs of society per se which justifies conscription, but a morally commendable *form* of society.
5 Jonas's published remarks on abortion are, to my knowledge, restricted to a digression in a discussion of negative eugenics (*PE*: 151–3). There Jonas claims in a footnote that abortion 'is always a violation of the most fundamental of all rights, the right to live' (*PE*: 148). This remark is later qualified, however, when he states that 'our moral sense is willing to consider [abortion permissible] at sufficiently

early stages' (*PE*: 152). The reasons for this exception are not discussed at any length: Jonas merely suggests that 'maternal disinclination' provides an insufficient justification for having an abortion, whereas 'the anticipation of grave deformity' in the infant represents 'the most defensible if not outright compelling ground' (*PE*: 152).

6   The difference between 'active' and 'passive' euthanasia aligns, to some extent, with the difference between killing and letting die – except that passive euthanasia involves, as we shall see, the removal of life support systems or cessation of treatment. This is clearly an act, one which brings about death, and one might ask what then distinguishes it from, say, letting go of someone hanging over a precipice (a removal of life support in a very literal sense). The answer is that the latter brings about death through causes *external* to the body. By contrast, the removal of medical life support brings about death through causes which are *internal* to the body, and which were only temporarily arrested through active intervention.

7   The most plausible generic exception is language, which holds a unique status due to its hermeneutic centrality: it is the foremost ground on which understanding occurs, allowing subsequent learning to take place. The first language learnt, one's mother tongue, is incomparable even with additional languages that are learnt on the basis of the first. Even so, in grammar language possesses an internal logic that allows its usage to be refined or mastered, including in opposition to the way one was taught, thus preserving the fundamental symmetry of a shared ground. One might also note that a first language can be supplanted by a second through exclusive use of the latter, even though the second is initially learnt through the first.

8   A notable exception was the compulsory moral enhancement initially proposed by Julian Savulescu and Ingmar Persson (2008: 174), which they subsequently dropped in favour of a voluntary programme.

9   A polymath is of course capable of mastery in multiple domains, but this is remarkable precisely because they do so within the constraints of the human condition.

10 On this I suspect intuitions will diverge: some may find the threats to freedom compelling, while others might feel that these are overridden by the impulse to cure and prevent diseases.

Conclusion

1   Indeed, I owe this textual discovery to Christopher Groves, courtesy of his unpublished draft essay 'The Future Lives in Everything: Care, Conatus and *The Road*'.

2 The choice of words is not accidental: a quotation from Böhme also serves as one of the epigraphs of McCarthy's earlier novel *Blood Meridian*.
3 McCarthy's theology is here comparable to Jonas's myth of the God who, as immanent in the world, becomes in accordance with the events and acts of His Creation (*PL*: 275–7), and in whose omnipresence the past is eternally retained (*PU*: 185–7).

References

[AKT] Jonas, H. (1977), 'Acting, Knowing, Thinking: Gleanings from Hannah Arendt's Philosophical Work', *Social Research*, 44 (1): 25–43.

[APF] Jonas, H. (1965), *Augustin und das paulinische Freiheitsproblem: Eine philosophische Studie zum pelagianischen Streit*, 2nd edn, Göttingen: Vandenhoeck & Ruprecht.

[CBE] Jonas, H. (2001), 'Closer to the Bitter End', *Graduate Faculty Philosophy Journal*, 23 (1): 21–30.

[CG] Jonas, H. (1951), 'Comment on General System Theory', *Human Biology*, 23 (4): 328–35.

[CR] Jonas, H. (1992), 'The Consumer's Responsibility: A Commentary to Lenk', in A. Øfsti (ed.), *Ecology and Ethics: A Report From the Melbu Conference, 18–23 July 1990*, 215–18, Trondheim: Nordland Akademi for Kunst og Vitenskap.

[DPL] Jonas, H. (1994), *Das Prinzip Leben: Ansätze zu einer philosophischen Biologie*, trans. by H. Jonas and K. Dockhorn, Frankfurt am Main: Suhrkamp.

[EBA] Jonas, H. (1985), 'Ethics and Biogenetic Art', *Social Research*, 52 (3): 491–504.

[EDM] Jonas, H. (n.d.), 'Essay on Dualism and Monism', Philosophisches Archiv der Universität Konstanz, Konstanz: HJ 2-7a-3.

[EH] Jonas, H. (2001), 'Edmund Husserl and the Ontological Question', *Etudes Phénoménologiques*, 17 (33/34): 5–20.

[FOS] Jonas, H. (n.d.), 'The Female of the Species', Philosophisches Archiv der Universität Konstanz, Konstanz: HJ 4-9-5.

[FSI] Jonas, H. (1976), 'Freedom of Scientific Inquiry and the Public Interest', *Hastings Center Report*, 6 (4): 15–17.

[FWT] Jonas, H. (2005), *Fatalismus wäre Todsünde: Gespräche über Ethik und Mitverantwortung im dritten Jahrtausend*, ed. D. Böhler, Münster: LIT.

[GR] Jonas, H. (2001), *The Gnostic Religion: The Message of the Alien God and the Beginnings of Christianity*, 3rd edn, Boston, MA: Beacon Press.

[HA] Jonas, H. (1976), 'Hannah Arendt: 1906–1975', *Social Research*, 43 (1): 3–5.

[HBT] Jonas, H. (1967). 'Heidegger: *Being and Time*, Class Notes (Spring, 1967)', Philosophisches Archiv der Universität Konstanz, Konstanz: HJ 1-4-6.

[HRR] Jonas, H. (1990), 'Heidegger's Resoluteness and Resolve', in G. Neske and E. Kettering (eds), *Martin Heidegger and National Socialism: Questions and Answers*, 197–203, trans. by L. Harries, New York City, NY: Paragon House.

[IHJ] Jonas, H. and H. Scodel (2003), 'An Interview with Professor Hans Jonas', *Social Research*, 70 (2): 339–68.

[IR] Jonas, H. (1984), *The Imperative of Responsibility: In Search of an Ethics for the Technological Age*, trans. by H. Jonas and D. Herr, Chicago, IL: Chicago University Press.

[KG] Jonas, H. (1959), 'Kurt Goldstein and Philosophy', *The American Journal of Psychoanalysis*, 19 (2): 161–4.

[KM] Jonas, H. (1929), 'Karl Mannheims Soziologie des Geistes', *Schriften der Deutschen Gesellschaft für Soziologie*, 6 (1): 111–4.

[M] Jonas, H. (2008), *Memoirs*, ed. by C. Wiese, trans. by K. Winston, Lebanon, NH: Brandeis University Press.

[MM] Jonas, H. (1996), *Mortality and Morality: A Search for the Good After Auschwitz*, ed. by L. Vogel, Evanston, IL: Northwestern University Press.

[OF] Jonas, H. (2016), *Organism and Freedom: An Essay in Philosophical Biology*, ed. by J. O. Beckers and F. Preußger, Freiburg im Breisgau: Rombach. Available at: http://hans-jonas-edition.de/?p=142 (accessed 24 November 2016).

Note: each reference to *Organism and Freedom* includes a Latin numeral referring to the chapter (e.g. *OF* IV), as these have not been synthesized into a single manuscript.

[OFL] Jonas, H. (1943), 'On the Firing Line', *The Hebrew Union College Monthly*, January, 13–4.

[OG] Jonas, H. (1984), 'Ontological Grounding of a Political Ethics: On the Metaphysics of Commitment to the Future of Man', *Graduate Faculty Philosophy Journal*, 10 (1): 47–61.

[OPW] Jonas, H. (2007), 'Our Part in this War: A Word to Jewish Men', in C. Wiese, *The Life and Thought of Hans Jonas: Jewish Dimensions*, 167–76, trans. by J. Grossman and C. Wiese, Lebanon, NH: Brandeis University Press.

[OR] Jonas, H. (n.d.), 'Oral Remarks at Bicentennial Conference', Philosophisches Archiv der Universität Konstanz, Konstanz: HJ 17-12-2.

[OS] Jonas, H. (2012), 'On Suffering', *La Rassegna Mensile di Israel*, 78 (1/2): 22–30.

[PE] Jonas, H. ((1974) 2010), *Philosophical Essays: From Ancient Creed to Technological Man*, New York City, NY: Atropos Press.

[PF] Jonas, H. (2010), *Problemi di libertà*, ed. by E. Spinelli and A. Michelis, Turin: Nino Aragno Editore.

[PL] Jonas, H. ((1966) 2001), *The Phenomenon of Life: Toward a Philosophical Biology*, Evanston, IL: Northwestern University Press.

[PU] Jonas, H. (1992), *Philosophische Untersuchungen und metaphysische Vermutungen*, Frankfurt am Main: Insel.

[PV] Jonas, H. (1979), *Das Prinzip Verantwortung: Versuch einer Ethik für die technologische Zivilisation*, Frankfurt am Main: Suhrkamp.

[RD] Jonas, H. (1978), 'The Right to Die', *Hastings Center Report*, 8 (4): 31–6.

[SB] Jonas, H. (1978), 'Straddling the Boundaries of Theory and Practice: Recombinant DNA Research as a Case of Action in the Process of Inquiry', in J. Richards (ed.),

Recombinant DNA: Science, Ethics, and Politics, 253-71, New York City, NY: Academic Press.
[*SE*] Jonas, H. (n.d.), 'Science and Ethics', Philosophisches Archiv der Universität Konstanz, Konstanz: HJ 1-10-5e.
[*TH*] Jonas, H. (1972), 'Testimony Before Subcommittee on Health, United States Senate', Library of Congress, Washington DC: 73-191-0.
[*THSR*] Jonas, H. (1977), 'Testimony Before Subcommittee on Health and Scientific Research, United States Senate', Philosophisches Archiv der Universität Konstanz, Konstanz: HJ 7-1-1.
[*TME*] Jonas, H. (1985), *Technik, Medizin und Ethik: Zur Praxis des Prinzips Verantwortung*, Frankfurt am Main: Suhrkamp.
[*TPT*] Jonas, H. (1979), 'Toward a Philosophy of Technology', *Hastings Center Report*, 9 (1): 34-43.
[*TSE*] Jonas, H. (1982), 'Technology as a Subject for Ethics', *Social Research*, 49 (4): 891-8.
[*WGM*] Jonas, H. (n.d.), 'What Does "Good" Mean in "Good Physician"?', Philosophisches Archiv der Universität Konstanz, Konstanz: HJ 1-2-12.
[*WPE*] Jonas, H. (2002), '*Wissenschaft* as Personal Experience', *Hastings Center Report*, 32 (4): 27-35.
Anders, G. (1956), *Die Antiquiertheit des Menschen: Über die Seele im Zeitalter der zweiten industriellen Revolution, Vol. 1*, Munich: C. H. Beck.
Apel, K.-O. (1996), *Selected Essays, Vol. 2: Ethics and the Theory of Rationality*, ed. by E. Mendieta, trans. by W. Brown, Atlantic Highlands, NJ: Humanities Press.
Arendt, H. (1958), *The Human Condition*, Chicago, IL: University of Chicago Press.
Arendt, H. (1979), 'On Hannah Arendt', in M. A. Hill (ed.), *Hannah Arendt and the Recovery of the Public World*, 301-40, New York City, NY: St Martin's.
Aristotle (1984a), 'Generation of Animals', trans. by A. Platt, in J. Barnes (ed.) *Complete Works, Vol. 1*, 1111-218, Princeton, NJ: Princeton University Press.
Aristotle (1984b), 'Metaphysics', trans. by W. D. Ross, in J. Barnes (ed.), *Complete Works, Vol. 2*, 1552-728, Princeton, NJ: Princeton University Press.
Aristotle (1984c), 'Movement of Animals', trans. by A. S. L. Farquharson, in J. Barnes (ed.), *Complete Works, Vol. 1*, 1087-96, Princeton, NJ: Princeton University Press.
Aristotle (1984d), 'Nicomachean Ethics', trans. by B. Jowett, in J. Barnes (ed.), *Complete Works, Vol. 2*, 1729-867, Princeton, NJ: Princeton University Press.
Aristotle (1984e), 'On the Soul', trans. by J. A. Smith, in J. Barnes (ed.), *Complete Works, Vol. 1*, 641-92, Princeton, NJ: Princeton University Press.
Aristotle (1984f), 'Parts of Animals', trans. by W. Ogle, in J. Barnes (ed.), *Complete Works, Vol. 1*, 994-1086, Princeton, NJ: Princeton University Press.
Aristotle (1984g), 'Physics', trans. by R. P. Hardie and R. K. Gaye, in J. Barnes (ed.), *Complete Works, Vol. 1*, 315-446, Princeton, NJ: Princeton University Press.
Aristotle (1984h), 'Politics', trans. by W. D. Ross and J. O. Urmson, in J. Barnes (ed.), *Complete Works, Vol. 2*, 1986-2129, Princeton, NJ: Princeton University Press.

Aristotle (1984i), 'Posterior Analytics', trans. by J. Barnes, in J. Barnes (ed.), *Complete Works, Vol. 1*, 114–66, Princeton, NJ: Princeton University Press.

Attfield, R. (1991), *The Ethics of Environmental Concern*, 2nd edn, Athens, GA: University of Georgia Press.

Aubert, M., P. Setiawan, A. A. Oktaviana, A. Brumm, P. H. Sulistyarto, E. W. Saptomo, B. Istiawan, T. A. Ma'rifat, V. N. Wahyuono, F. T. Atmoko, J.-X. Zhao, J. Huntley, P. S. C. Taçon, D. L. Howard and H. E. A. Brand (2018), 'Palaeolithic Cave Art in Borneo', *Nature*, 564: 254–7.

Bacon, F. ((1605/1627) 1906), *The Advancement of Learning and New Atlantis*, Oxford: Oxford University Press.

Bacon, F. (1984), *Valerius Terminus: Von der Interpretation der Natur*, trans. by F. Träger, Würzburg: Königshausen and Neumann.

Bacon, F. ((1620) 2000), *The New Organum*, ed. by L. Jardine and M. Silverthorne, Cambridge: Cambridge University Press.

Barbaras, R. (2008), 'Life, Movement, and Desire', *Research in Phenomenology*, 38 (1): 3–17.

Barker, G. (2015), *Beyond Biofatalism: Human Nature for an Evolving World*, New York City, NY: Columbia University Press.

Bernstein, R. J. (1995), 'Rethinking Responsibility', *Hastings Center Report*, 25 (7): 13–20.

Bertalanffy, L. (1952), *Problems of Life: An Evaluation of Modern Biological and Scientific Thought*, New York City, NY: Harper & Brothers.

Blande, J. D. and R. Glinwood, eds (2016), *Deciphering Chemical Language of Plant Communication*, Cham: Springer.

Bloch, E. (1986), *The Principle of Hope*, trans. by N. Plaice, S. Plaice and P. Knight, Cambridge, MA: MIT Press.

Bookchin, M. (1982), *The Ecology of Freedom: The Emergence and Dissolution of Hierarchy*, Palo Alto, CA: Cheshire Books.

Bostrom, N. (2008), 'Letter From Utopia', *Studies in Ethics, Law, and Technology*, 2 (1): 1–7.

Callicott, J. B. (1984), 'Non-Anthropocentric Value Theory and Environmental Ethics', *American Philosophical Quarterly*, 21 (4): 299–309.

Callicott, J. B. (1989), *In Defence of the Land Ethic: Essays in Environmental Philosophy*, Albany, NY: State University of New York Press.

Cassirer, E. (1944), *An Essay on Man: An Introduction to a Philosophy of Human Culture*, New Haven, CT: Yale University Press.

Collingwood, R. G. (1945), *The Idea of Nature*, Oxford: Oxford University Press.

Descartes, R. (1968), *Discourse on Method and the Meditations*, trans. by F. E. Sutcliffe, London: Penguin.

Descartes, R. (1972), *Treatise of Man*, trans. by T. S. Hall, Cambridge, MA: Harvard University Press.

Dinneen, N. (2014), 'Hans Jonas's Noble "Heuristics of Fear": Neither the Good Lie Nor the Terrible Truth', *Cosmos and History*, 10 (1): 1–21.

Dinneen, N. (2017), 'Ecological Scenario Planning and the Question of the Best Regime in the Political Theory of Hans Jonas', *Environmental Politics*, 26 (5): 938–55.

Dostoevsky, F. (1992), *The Brothers Karamazov*, trans. by R. Pevear and L. Volokhonsky, London: Vintage.

Dreyfus, H. L. (1991), *Being-in-the-World: A Commentary on Heidegger's Being and Time, Division I*, Cambridge, MA: MIT Press.

Farrell, J. M. (2015), 'The Baby and the Bathwater: Hans Jonas's Recovery of Aristotelian Biological Concepts', *Journal of the British Society for Phenomenology*, 45 (3): 187–202.

Ferry, L. (1995), *The New Ecological Order*, trans. by C. Volk, Chicago, IL: Chicago University Press.

File, A. L., J. Klironomos, H. Maherali and S. A. Dudley (2012), 'Plant Kin Recognition Enhances Abundance of Symbiotic Microbial Partner', *PLoS One*, 7 (9): 1–10.

Franzini Tibaldeo, R. (2009), *La rivoluzione ontologica di Hans Jonas: Uno studio sulla genesi e il significato di "Organismo e libertà"*, Milano-Udine: Mimesis.

Frogneux, N. (2014), 'Some Paradoxes Linked to Risk Moderation', trans. by J. Cronin, in J.-S. Gordon and H. Burckhart (eds), *Global Ethics and Moral Responsibility: Hans Jonas and His Critics*, 73–91, Farnham: Ashgate.

Fukuyama, F. (2002), *Our Posthuman Future: Consequences of the Biotechnology Revolution*, New York City, NY: Farrar, Strauss & Giroux.

Furnari, M. G. (2006), 'From the Ontology of Temporality to the Ethics of the Future: Care and Responsibility in Hans Jonas', in L. V. Siegal (ed.), *Philosophy and Ethics: New Research*, 133–57, New York City, NY: Nova Science Publishers.

Gadamer, H.-G. (1976), *Philosophical Hermeneutics*, trans. by D. E. Linge, Berkeley, CA: University of California Press.

Gadamer, H.-G. (1996), *The Enigma of Health: The Art of Healing in a Scientific Age*, trans. by J. Gaiger and N. Walker, Cambridge: Polity.

Gadamer, H.-G. (2004), *Truth and Method*, 2nd edn, trans. by J. Weinsheimer and D. G. Marshall, London: Continuum.

Galileo (2008), *The Essential Galileo*, ed. and trans. by M. A. Finocchiaro, Indianapolis, IN: Hackett.

Gehlen, A. (1988), *Man: His Nature and Place in the World*, trans. by C. McMillan and K. Pillemer, New York City, NY: Columbia University Press.

Genesis 1–4:26, *Holy Bible*, King James Version.

Gorz, A. (1980), *Ecology as Politics*, trans. by P. Vigderman and J. Cloud, Montréal, QC: Black Rose Press.

Habermas, J. (1988), *On the Logic of the Social Sciences*, trans. by S. W. Nicholsen and J. A. Stark, Cambridge, MA: MIT Press.

Habermas, J. (2003), *The Future of Human Nature*, trans. by W. Rehg, M. Pensky and H. Beister, Cambridge: Polity.

Hannah Arendt (2012), [Film] Dir. Margarethe von Trotta, Germany/Luxembourg/France: Heimatfilm, Bayerische Rundfunk and Westdeutsche Rundfunk.

Harris, J. (1975), 'The Survival Lottery', *Philosophy*, 50 (191): 81–7.
Harris, J. (2007), *Enhancing Evolution: The Ethical Case for Making Better People*, Princeton, NJ: Princeton University Press.
Hauskeller, M. (2007), *Biotechnology and the Integrity of Life: Taking Public Fears Seriously*, Farnham: Ashgate.
Hauskeller, M. (2015), 'The Ontological Ethics of Hans Jonas', in D. Meacham (ed.), *Medicine and Society: New Perspectives in Continental Philosophy*, 39–55, Dordrecht: Springer.
Hegel, G. W. F. (1975), *Aesthetics: Lectures on Fine Art, Vol. 1*, trans. by T. M. Knox, Oxford: Oxford University Press.
Heidegger, M. (1966), *Discourse on Thinking*, trans. by J. M. Anderson and E. H. Freund, New York City, NY: Harper & Row.
Heidegger, M. (1977), *Basic Writings*, ed. by D. F. Krell, London: Routledge.
Heidegger, M. (1985), *History of the Concept of Time: Prologomena*, trans. by T. Kisiel, Bloomington, IN: Indiana University Press.
Heidegger, M. (1990), '*Der Spiegel* Interview with Martin Heidegger', in G. Neske and E. Kettering (eds), *Martin Heidegger and National Socialism: Questions and Answers*, 41–66, trans. by L. Harries, New York City, NY: Paragon House.
Heidegger, M. (1995), *The Fundamental Concepts of Metaphysics: World, Finitude, Solitude*, trans. by W. McNeill and N. Walker, Bloomington, IN: Indiana University Press.
Heidegger, M. (1997a), *Kant and the Problem of Metaphysics*, 5th edn, trans. by R. Taft, Bloomington, IN: Indiana University Press.
Heidegger, M. (1997b), *Plato's Sophist*, trans. by R. Rojcewicz and A. Schuwer, Bloomington, IN: Indiana University Press.
Heidegger, M. (2000), *Introduction to Metaphysics*, trans. by G. Fried and R. Polt, New Haven, CT: Yale University Press.
Heidegger, M. (2001), *Zollikon Seminars: Protocols–Conversations–Letters*, ed. by M. Boss, trans. by F. Mayr and R. Askay, Evanston, IL: Northwestern University Press.
Heidegger, M. (2002), *Supplements: From the Earliest Essays to Being and Time and Beyond*, trans. by J. van Buren, New York City, NY: State University of New York Press.
Heidegger, M. (2009), *Basic Concepts of Aristotelian Philosophy*, trans. by R. D. Metcalf and M. B. Tanzer, Bloomington, IN: Indiana University Press.
Heidegger, M. (2010a), *Being and Time*, trans. by J. Stambaugh and D. J. Schmidt, Albany, NY: State University of New York Press.
Heidegger, M. (2010b), *Country Path Conversations*, trans. by B. W. Davis, Bloomington, IN: Indiana University Press.
Heidegger, M. (2010c), *Logic: The Question of Truth*, trans. by T. Sheehan, Bloomington, IN: Indiana University Press.
Heidegger, M. (2016), *Ponderings II–VI: Black Notebooks 1931–1938*, trans. by R. Rojcewicz, Bloomington, IN: Indiana University Press.

Heidegger, M. and E. Fink (1993), *Heraclitus Seminar*, trans. by C. H. Seibert, Evanston, IL: Northwestern University Press.

Herrmann, F.-W. v. (2016), 'The Role of Martin Heidegger's *Notebooks* within the Context of His Oeuvre', in I. Farin and J. Malpas (eds), *Reading Heidegger's Black Notebooks 1931–1941*, 89–94, Cambridge, MA: MIT Press.

Hobbes, T. ((1651) 1914), *Leviathan*, London: J. M. Dent & Sons.

Hösle, V. (2001), 'Ontology and Ethics in Hans Jonas', *Graduate Faculty Philosophy Journal*, 23 (1): 31–50.

Hottois, G. (1993), 'Introduction', in G. Hottois (ed.), *Aux Fondements D'une Éthique Contemporaine: H. Jonas et H.T. Engelhardt*, 11–31, Paris: Librarie Philosophique J. Vrin.

Hottois, G. (1996), *Entre Symboles & Technosciences: Un Itinéraire Philosophique*, Seyssel: Editions Champ Vallon.

Hume, D. ((1739/1740) 1969), *A Treatise of Human Nature*, London: Penguin.

Hume, D. (1975), *Enquiries Concerning Human Understanding and Concerning the Principles of Morals*, 3rd edn, Oxford: Clarendon Press.

Husserl, E. (1970), *The Crisis of European Sciences and Transcendental Phenomenology: An Introduction to Phenomenological Philosophy*, trans. by D. Carr, Evanston, IL: Northwestern University Press.

Husserl, E. (1999), *Cartesian Meditations: An Introduction to Phenomenology*, trans. by D. Cairns, Dordrecht: Kluwer Academic Publishers.

Ihde, D. (1979), *Technics and Praxis*, Dordrecht: D. Reidel Publishing.

IPCC (2013), 'Summary for Policymakers', in *Climate Change 2013: The Physical Science Basis. Contribution of Working Group I to the Fifth Assessment Report of the Intergovernmental Panel on Climate Change*, Cambridge: IPCC.

IPCC (2014), 'Summary for Policymakers', in *Climate Change 2014: Impacts, Adaptation, and Vulnerability. Part A: Global and Sectoral Aspects. Contribution of Working Group II to the Fifth Assessment Report of the Intergovernmental Panel on Climate Change*, Cambridge: IPCC.

Jesus, P. d. (2016), 'Autopoietic Enactivism, Phenomenology and the Deep Continuity Between Life and Mind', *Phenomenology and the Cognitive Sciences*, 15 (2): 265–89.

Johnson, A. R. (2012), 'Avoiding Environmental Catastrophes: Varieties of Principled Precaution', *Ecology and Society*, 17 (3): 1–13.

Johnson, A. T. (2014), 'Is Organic Life "Existential"? Reflections on the Biophenomenologies of Hans Jonas and Early Heidegger', *Environmental Philosophy*, 11 (2): 253–77.

Jonsen, A. R. (1998), *The Birth of Bioethics*, Oxford: Oxford University Press.

Kampowski, S. (2013), *A Greater Freedom: Biotechnology, Love, and Human Destiny (in Dialogue with Hans Jonas and Jürgen Habermas)*, Eugene, OR: Pickwick Publications.

Kant, I. (1992), *Lectures on Logic*, ed. and trans. by J. M. Young, Cambridge: Cambridge University Press.

Kant, I. (1993), *Grounding for the Metaphysics of Morals*, 3rd edn, trans. by J. W. Ellington, Indianapolis, IN: Hackett.

Kant, I. (1996), *Metaphysics of Morals*, trans. by M. Gregor, Cambridge: Cambridge University Press.

Kant, I. ((1997) 2015), *Critique of Practical Reason*, trans. by M. Gregor, Cambridge: Cambridge University Press.

Karban, R. (2015), *Plant Sensing and Communication*, Chicago, IL: University of Chicago Press.

Karban, R., K. Shiojiri, S. Ishizaki, W. C. Wetzel and R. Y. Evans (2013), 'Kin Recognition Affects Plant Communication and Defence', *Proceedings of the Royal Society: Biological Sciences*, 280 (1756): 1–5.

Kass, L. R. (1995), 'Appreciating *The Phenomenon of Life*', *Hastings Center Report*, 25 (7): 3–12.

Kass, L. R. (2002), *Life, Liberty and the Defense of Dignity: The Challenge for Bioethics*, San Francisco, CA: Encounter Books.

Kisiel, T. J. (1993), *The Genesis of Heidegger's "Being and Time"*, Berkeley, CA: University of California Press.

Klostermann, V., ed. (1970), *Durchblicke: Martin Heidegger zum 80. Geburtstag*, Frankfurt am Main: Vittorio Klostermann.

Korsgaard, C. M. (1986), 'The Right to Lie: Kant on Dealing With Evil', *Philosophy and Public Affairs*, 15 (4): 325–49.

Krebs, A. (1999), *Ethics of Nature: A Map*, Berlin: Walter de Gruyter.

Kuhlmann, W. (1994), '"Prinzip Verantwortung" versus Diskursethik', in D. Böhler (ed.), *Ethik für die Zukunft: Im Diskurs mit Hans Jonas*, 277–302, Frankfurt am Main: C. H. Beck.

Kurasawa, F. (2007), *The Work of Global Justice: Human Rights as Practices*, Cambridge: Cambridge University Press.

Lapidot, E. (2017), 'Hans Jonas' Work on Gnosticism as Counterhistory', *Philosophical Readings*, 9 (1): 61–8.

Levy, D. J. (1987), *Political Order: Philosophical Anthropology, Modernity, and the Challenge of Ideology*, Baton Rouge, LA: Louisiana State University Press.

Levy, D. J. (2002), *Hans Jonas: The Integrity of Thinking*, Columbia, MO: University of Missouri Press.

Lewis, J. (1974), 'Foreword', in J. Lewis (ed.), *Beyond Chance and Necessity: A Critical Inquiry into Professor Jacques Monod's Chance and Necessity*, ix–xi, London: The Garnstone Press.

Lindberg, S. (2005), 'Hans Jonas' Theory of Life in the Face of Responsibility', in K.-H. Lembeck, K. Mertens and E. W. Orth (eds), *Phänomenologischen Forschungen*, 175–91, Hamburg: Felix Meiner Verlag.

MacIntyre, A. (2002), 'On Not Having the Last Word: Thoughts on Our Debts to Gadamer', in J. Malpas, U. Arnswald and J. Kertscher (eds), *Gadamer's Century: Essays in Honor of Hans-Georg Gadamer*, 157–72, Cambridge, MA: MIT Press.

MacIntyre, A. (2007), *After Virtue: A Study in Moral Theory*, 3rd edn, Notre Dame, Indiana, IN: University of Notre Dame Press.
Mannheim, K. ([1936] 1991), *Ideology and Utopia: An Introduction to the Sociology of Knowledge*, trans. by L. Wirth, London: Routledge.
Mayr, E. (1988), *Toward a New Philosophy of Biology: Observations of an Evolutionist*, Cambridge, MA: Harvard University Press.
McCarthy, C. (2006), *The Road*, London: Picador.
McKenny, G. P. (1997), *To Relieve the Human Condition: Bioethics, Technology, and the Body*, Albany, NY: State University of New York Press.
Melle, U. (1998), 'Responsibility and the Crisis of Technological Civilisation: A Husserlian Meditation on Hans Jonas', *Human Studies*, 21 (4): 329–45.
Morris, T. (2013), *Hans Jonas's Ethic of Responsibility: From Ontology to Ecology*, Albany, NY: State University of New York Press.
Moss, C. (2000), *Elephant Memories: Thirteen Years in the Life of an Elephant Family*, Chicago, IL: University of Chicago Press.
Moss, L. (2003), *What Genes Can't Do*, Cambridge, MA: MIT Press.
Moss, L. (2005), 'Darwinism, Dualism, and Biological Agency', in V. Hösle and C. Illies (eds), *Darwinism and Philosophy*, 349–63, Notre Dame, IN: University of Notre Dame Press.
Nietzsche, F. ((1973) 1990), *Beyond Good and Evil: Prelude to a Philosophy of the Future*, trans. by R. J. Hollingdale, London: Penguin.
Nietzsche, F. (2017), *The Will to Power: Selections from the Notebooks of the 1880s*, trans. by R. K. Hill and M. A. Scarpitti, London: Penguin.
Paolo, E. A. d. (2006), 'Autopoiesis, Adaptivity, Teleology, Agency', *Phenomenology and the Cognitive Sciences*, 4 (4): 429–52.
Persson, I. and J. Savulescu (2008), 'The Perils of Cognitive Enhancement and the Urgent Imperative to Enhance the Moral Character of Humanity', *Journal of Applied Philosophy*, 25 (3): 162–77.
Persson, I. and J. Savulescu (2012), *Unfit for the Future: The Need for Moral Enhancement*, Oxford: Oxford University Press.
Persson, I. and J. Savulescu (2015), 'The Art of Misunderstanding Moral Bioenhancement: Two Cases', *Cambridge Quarterly of Healthcare Ethics*, 24 (1): 48–57.
Pettit, P. (1997), *Republicanism: A Theory of Freedom and Government*, Oxford: Oxford University Press.
Pettit, P. (2012), *On the People's Terms: A Republican Theory and Model of Democracy*, Cambridge: Cambridge University Press.
Plato (1997), 'Critias', trans. by D. Clay, in J. M. Cooper (ed.), *Complete Works*, 1292–1306, Indianapolis, IN: Hackett.
Plessner, H. (1970), *Laughing and Crying: A Study of the Limits of Human Behavior*, 3rd edn, trans. by J. S. Churchill and M. Grene, Evanston, IL: Northwestern University Press.

Plessner, H. (2019), *Levels of Organic Life and the Human: An Introduction to Philosophical Anthropology*, trans. by M. Hyatt, New York City, NY: Fordham University Press.

Portmann, A. (1956), 'Letter', Philosophisches Archiv der Universität Konstanz, Konstanz: HJ 16-4-25.

Portmann, A. (1961), *Animals as Social Beings*, trans. by O. Coburn, London: Hutchinson.

Principe, L. M. and A. Weeks (1989), 'Jacob Boehme's Divine Substance *Salitter*: Its Nature, Origin, and Relationship to Seventeenth Century Scientific Theories', *The British Journal for the History of Science*, 22 (1): 53–61.

Rolston III, H. (1988), *Environmental Ethics: Duties to and Values in the Natural World*, Philadelphia, PA: Temple University Press.

Rowlands, M. (2012), *Can Animals Be Moral?*, Oxford: Oxford University Press.

Rutkoff, P. M. and W. B. Scott (1986), *New School: A History of the New School for Social Research*, New York City, NY: The Free Press.

Sandel, M. J. (1996), *Democracy's Discontent: America in Search of a Public Philosophy*, Cambridge, MA: The Belknap Press.

Savage-Rumbaugh, S., S. G. Shanker and T. J. Taylor (1998), *Apes, Language, and the Human Mind*, Oxford: Oxford University Press.

Scheler, M. (1961), *Man's Place in Nature*, trans. by H. Meyerhoff, New York City, NY: Noonday Press.

Schmidt, J. C. (2014), 'Ethics for the Technoscientific Age: On Hans Jonas' Argumentation and his Public Philosophy Beyond Disciplinary Boundaries', in J.-S. Gordon and H. Burckhart (eds), *Global Ethics and Moral Responsibility: Hans Jonas and His Critics*, 147–70, Farnham: Ashgate.

Schütze, C. (1995), 'The Political and Intellectual Influence of Hans Jonas', *Hastings Center Report*, 25 (7): 40–3.

Steffen, W., J. Rockström, K. Richardson, T. M. Lenton, C. Folke, D. Liverman, C. P. Summerhayes, A. D. Barnosky, S. E. Cornell, M. Crucifix, J. F. Donges, I. Fetzer, S. J. Lade, M. Scheffer, R. Winkelmann and H. J. Schellnhuber (2018), 'Trajectories of the Earth System in the Anthropocene', *PNAS*, 115 (33): 8252–9.

Sugiyama, Y. (1985), 'The Brush-Stick of Chimpanzees Found in South-West Cameroon and Their Cultural Characteristics', *Primates*, 26 (4): 361–74.

Taylor, P. W. ((1986) 2011), *Respect for Nature: A Theory of Environmental Ethics*, Princeton, NJ: Princeton University Press.

Thompson, E. (2004), 'Life and Mind: From Autopoiesis to Neurophenomenology. A Tribute to Francisco Varela', *Phenomenology and the Cognitive Sciences*, 3 (4): 381–98.

Tirosh-Samuelson, H. and C. Wiese, eds (2008), *The Legacy of Hans Jonas: Judaism and the Phenomenon of Life*, Leiden: Brill.

United Nations (1982), *World Charter for Nature: General Assembly Resolution 37/7*, 28th October.

United Nations (1992), *Report of the United Nations Conference on Environment and Development*, 12th August.

Villalobos, M. and D. Ward (2016), 'Lived Experience and Cognitive Science: Reappraising Enactivism's Jonasian Turn', *Constructivist Foundations*, 11 (2): 204–12.

Voegelin, E. (1952), *The New Science of Politics: An Introduction*, Chicago, IL: The University of Chicago Press.

Vogel, L. (1995), 'Does Environmental Ethics Need a Metaphysical Grounding?', *Hastings Center Report*, 25 (7): 30–9.

Vorstenbosch, J. (1993), 'The Concept of Integrity: Its Significance for the Discussion on Biotechnology and Animals', *Livestock Science*, 36 (1): 109–12.

Weber, A. and F. J. Varela (2002) 'Life After Kant: Natural Purposes and the Autopoietic Foundations of Biological Individuality', *Phenomenology and the Cognitive Sciences*, 1 (2): 97–125.

Weisskopf, W. A. (2014), 'Moral Responsibility for the Preservation of Humankind', in J.-S. Gordon and H. Burckhart (eds), *Global Ethics and Moral Responsibility: Hans Jonas and His Critics*, 23–40, Farnham: Ashgate.

Wennemann, D. J. (2013), *Posthuman Personhood*, Lanham, MD: University Press of America.

Whitehead, A. N. (1920), *The Concept of Nature*, Cambridge: Cambridge University Press.

Whiteside, K. H. (2006), *Precautionary Politics: Principle and Practice in Confronting Environmental Risk*, Cambridge, MA: MIT Press.

Wolin, R. ((2001) 2015), *Heidegger's Children: Hannah Arendt, Karl Löwith, Hans Jonas, and Herbert Marcuse*, Princeton, NJ: Princeton University Press.

Index

animal symbolicum 80–1, 130; *see also* symbol
animals 21–3, 42–3, 65–7, 70–9, 89–90, 127–8, 140–4
anthropocentrism 96, 118–19, 136–44
antiquity 2, 9–13, 167–72; *see also* Greco-Roman antiquity
Arendt, Hannah 4, 137, 167, 172, 186, 217
Aristotle 4, 209
   biology 21, 48–50, 65–6, 80, 102–3, 148
   ethics 25, 101, 169–70
   metaphysics 20–2, 89, 114
axiology 95–100, 104–9, 113–16; *see also* value
   objectivist 199–202
   relativist 110

Bacon, (Sir) Francis 30–3, 154–6, 186, 199–200
behaviour 47–9, 54–60, 96–8
behavioural modernity 85–7, 93, 213
Being 12–13, 18–24, 36–9, 60–6, 92–4, 107–9, 114–16, 142–4, 202, 210
   non- 115–16
Being and Time 13–14, 39–43, 80, 132, 158
being-in-the-world 42–6, 48, 56–60, 78; *see also* world (*Welt*)
being-toward-death 15–17, 59–60, 84, 196; *see also* death
biocentrism 96
   axiological 96–100
   ethical 136–44
bioethics 173–98
biosphere 100–8, 118, 137–9, 210
biotechnology 33–5, 185–98
Bloch, Ernst 155, 217
body 22, 38–49, 58, 73–4, 81, 182–5, 212
Bultmann, Rudolf 4, 210

call (*Gewissensruf*) 125–7, 132, 157–8; *see also* responsibility, call of
capitalism 27–30, 154–6
care (*Sörge*) 42–8, 58–60, 62, 67–70, 76–7, 96–100, 124–8, 134, 136–8, 158, 168, 217
categorical imperative 121–2, 130–1, 157; *see also* deontology
Christianity 1, 9–10, 16–18, 21, 30–2, 118–19, 124, 200–1; *see also* Judeo-Christianity
citizenship 12, 134–5, 156–9, 166–71
climate change 33–4, 120, 133–5, 203–4, 216; *see also* ecological crisis
communism 154–5, 159, 217; *see also* Marxist-Leninism
concern (*Fürsorgen*) 14–5, 41, 45, 70, 76–7, 85, 124–5
   self- 48, 55, 57–8, 61–2, 92
consciousness 38–41, 66, 73, 93; *see also* mind
   historical 111–12
cosmology 9–11, 119, 204–5
cosmos 10–12, 18, 21, 66, 103
culture 80, 85–7, 110–12

Darwin, Charles 47, 65–6, 102–3
Dasein 14–6, 40–4, 57–60, 77, 81, 92, 212
death 11–12, 30, 61–2, 84–5, 181–5, 204–6, 219; *see also* mortality
democracy 156, 160, 172
deontology 109, 121–2, 136; *see also* categorial imperative
Descartes, René 22–3, 31, 37–40, 42, 45–6, 51–2
differentia specifica 79–87
dignity 23–4, 137–43, 175
   of ends 138, 140, 184
   human 140–1, 174–80, 183–6, 190, 218

non-personal 140–1, 143, 184–6, 190, 193–4, 197–8
personal 140–1, 143, 171, 185, 197
dualism 10–13, 16, 22–3, 37–48, 61–3, 182
duty 120–3, 128, 131–8, 179–81, 215

ecological crisis 6–7, 33–5, 159–65, 171–2; *see also* climate change
ecology 103–5, 154–6, 207
eidos 88–93, 130, 175
emotion 72–8, 113–15, 126–8, 135–6
end-in-itself 96, 100–6, 140–1
environment 137–8, 152–5, 215
Umwelt 41–3, 54–9, 67–70, 75, 91
environmental ethics 6, 96, 99–101, 123, 217
eschatology 30–3, 115, 154–6, 160–1
ethics 6–7, 28–9, 33–6, 95–6, 113–44, 151, 173–98, 200–2, 214–16; *see also* morality
eugenics 190–1
euthanasia 184, 219
existence (*Existenz*) 14–6, 44, 62, 76–7, 91–2, 111, 128, 147–8, 212
existential analytic 40–4, 58–60, 81, 92–3
existential freedom 140, 148, 186–8; *see also* freedom
existential phenomenology 39–40, 70–1, 212; *see also* phenomenology
existentialism 13–9, 39–40, 44, 93, 210–11

freedom 48, 59, 68, 75–6, 90, 145–8, 163–71, 188–90, 194, 205–6, 215; *see also* existential freedom
future generations 119–23, 132–3, 215

Gadamer, Hans-Georg 111–14, 192–4
Genesis 10, 32, 114, 119, 123
genetic engineering 34–5, 186–90, 196–8, 211
Gnostic religion 4, 10–18, 30–1, 210
gnosticism 2, 7–8, 11–19, 23–4, 30–3, 38–40, 93–4, 114–15, 117–19, 160–1, 166, 182, 194, 199–200, 211, 218
God 9–12, 29–30, 114, 119, 140, 205–6, 220

good 11–12, 23, 118, 128–30, 205–6
common 105, 166–8, 170
-in-itself 108–9
of its kind 101–3
of its own 95, 98–9
moral 128–30
objective 16–17, 107–9, 113–16, 127–8
subjective 99, 107, 125
great chain of Being 21, 65; *see also scala naturae*
Greco-Roman antiquity 2, 12, 17–18, 110, 113–14; *see also* antiquity
health 25, 99, 191–5
healthcare 174, 191–4; *see also* medicine
Heidegger, Martin 4–5, 13–18, 28–30, 39–45, 56–9, 70, 76–7, 80–1, 92–3, 111, 124–5, 132, 158, 160, 163–4, 202–3, 209–14, 217
hermeneutics 16, 111–12, 219
heuristic of fear 148–50, 161–5, 174–5, 186, 206
human beings 7–8, 10–11, 15–17, 21–4, 35–6, 43–4, 66, 79–94, 110–13, 118–22, 128–32, 136–44, 148, 150–2, 156, 163, 175–85, 191, 202–3, 205–6, 215–16
human condition 25, 118, 134, 137, 168, 185–6, 194–8, 202, 219
human enhancement 185–98; *see also* transhumanism
human nature 19, 31, 113
Hume, David 107–9, 113–14, 126, 136
Husserl, Edmund 4, 39–41, 45

idea of Man 93, 130–1, 138–9, 143–4, 149, 157, 173–5, 191, 197–8, 206–7, 216; *see also* image of Man
idealism 23, 38–40, 43–4
image 86–93, 130, 175, 191, 214
of Man 93, 130, 191, 216 (*see also* idea of Man)
imago Dei 119, 140, 163, 206
immortality 32, 186, 194–6
integrity 118, 137–43, 184–5
is–ought gap 107–8

Jews 1, 3–6, 209
Judaism 10

Judeo-Christianity 2, 32, 110, 113–14, 119–20, 123–4, 200–1, 205; *see also* Christianity

Kant, Immanuel 38, 79–80, 120–3, 136, 140–1, 170–1, 178–9, 183–4, 215

language 85–6, 90, 110–12, 214, 219
Lebensphilosophie 62, 163–5
life 20–4, 37–67, 92–4, 99–111, 114, 126–9, 136–44, 147, 163–5, 181–5, 194–6, 201–7, 212–13
  animal (*see* animals)
  human (*see* human beings)
  vegetative (*see* plants)
living beings 20–3, 42–63, 96–100, 114–16, 125–8, 136–44, 181–5, 164, 201; *see also* organism
locomotion 21, 65, 68–76, 78

McCarthy, Cormac 203–6, 220
MacIntyre, Alasdair 113
Marxist-Leninism 154–6, 159, 217; *see also* Soviet Union
materialism 18–23, 31, 37–9, 46–7, 63, 138–9, 194
mathematics 21–2, 26, 28–30, 194
medicine 173–82, 184, 191–3, 219; *see also* healthcare
metabolism 49–54, 56–60, 67–8, 90, 182–3, 212–13
mind 22–3, 30–2, 38–9, 93, 182, 185, 195, 212; *see also* sentience
modernity 2, 7–8, 17–39, 43–4, 115–18, 154, 164–5, 167, 192–4, 199–200
moral agency 124–5, 137, 178, 216
moral being 115, 130–1, 141–3, 175, 202, 216
moral considerability 96, 124–7, 129, 142–3
moral law 121–2, 126–7, 136
moral responsibility 124–5, 128–9, 166, 203, 215
moral significance 129, 136–8, 142–3, 175
morality 29–30, 113–14, 117–45, 168, 174–6, 200–2; *see also* ethics
  capacity for 129–30, 132, 140–1, 143, 156, 175, 183, 206, 215–16
mortality 14–15, 76, 84–5, 194–6; *see also* death

movement 20, 26, 49–51, 55–6, 68–73, 91, 110, 117, 148, 212–13

natality 77–8, 194–5
natural world 7, 10–11, 15–16, 24, 30, 137
Nazism 1, 4, 6, 158
Nietzsche, Friedrich 17, 24
nihilism 7, 11, 16–19, 24, 99, 106, 108–9, 115–16
nisus of Being 60–3, 107–9, 142–3, 206, 213

organism 20, 45–63, 67–70, 73–7, 96–100, 105, 127–8, 139–41, 147, 182–5, 192–4; *see also* living beings

phenomenology 39–48, 58, 70–1, 125–7; *see also* existential phenomenology
philosophical anthropology 79–92, 130, 146–8, 182, 214
plants 21–3, 65–73, 143–4, 213
polis 12, 17–18, 146, 148, 166–71, 194; *see also res publica*
precautionary principle 150–3, 217
proprioception 73–4

republicanism 146–7, 165–72, 188–90, 198
res publica 12, 18, 148, 157–8, 166–7, 171; *see also polis*
responsibility 7, 123–6, 199–200, 215–16
  call of 126–7, 134, 141, 157–8, 207 (*see also* call (*Gewissensruf*))
  ethic of 123–44
  experience of 127, 135, 149, 168
  public 156–9, 166–9, 200
The Road 203–6, 219

Salitter 205–6
scala naturae 21, 24, 65–7; *see also* great chain of Being
science 19–33, 46–7, 57–8, 113, 138–9, 148–9, 173–6, 179–80, 192–3
self-organization 48–54, 56–7, 61–2, 98–100, 105, 213–14
sentience 61, 72–6, 97–8, 147, 206; *see also* consciousness

Soviet Union 154–6, 159–60, 217; *see also* communism
spatiality 49–50, 57–8, 68, 70–1, 74–5, 118, 175, 212
species 33–4, 86, 100–3, 113, 141–2
statesmanship 156–61, 166, 169
symbol 80–1, 86–93; *see also animal symbolicum*

technē 86, 118, 174, 192, 214
technology 2, 24–35, 83–4, 91–2, 117–18, 123, 132–3, 149–55, 164–5, 173–5, 193, 199–200, 209, 211, 217
teleology 20–3, 26, 46–56, 61–2, 95–100, 103, 114–16, 118–19, 128, 136, 140–1, 183, 213
telos 20, 32, 50, 61–2, 98–9, 113, 129–30, 139–42, 160–1, 190–4, 218
temporality 14, 58–9, 117, 120, 132, 196, 212
tradition 85, 109–13
   moral 113–16, 127–8, 170, 201–2
   Western 2–4, 118–19, 168–9
transcendence 3, 17–9, 24, 44, 56, 59, 90–1, 130, 133, 143–4, 203
transhumanism 185–97; *see also* human enhancement

utilitarianism 100, 120–1, 136–8, 174, 176–8, 181–3, 218
utopianism 7–8, 31–3, 117, 151–6, 160–1, 166, 171, 211, 217
   techno- 31, 154–5, 211

value 12–13, 17–19, 23–5, 30–1, 95–9, 106, 111–13, 116, 129, 143–4, 149–51; *see also* axiology
   instrumental 99–101, 105, 107
   intrinsic 99–101, 104–5
   objective 109–10, 114–16, 125
   species- 102–3
   subjective 109–10, 116
   systemic 104–5
virtue 113–14, 120, 134–6, 145, 167–70, 214–15
   ecological 167–70
Voegelin, Eric 31, 160–1, 163, 171–2

Whitehead, Alfred North 61–2
world (*Welt*) 14–16, 40–6, 56–7, 67, 77–8, 90–2, 110–11, 137, 193–5; *see also* being-in-the-world
   -openness 41–3, 56, 63, 66, 72, 80, 88, 94
   -poverty 43, 69–70

Zionism 3–6

www.ingramcontent.com/pod-product-compliance
Lightning Source LLC
Chambersburg PA
CBHW072147290426
44111CB00012B/1995